Volatile Organic Compounds in Environment

Special Issue Editor
Ki-Hyun Kim

MDPI • Basel • Beijing • Wuhan • Barcelona • Belgrade

MDPI

Special Issue Editor
Ki-Hyun Kim
Hanyang University
Korea

Editorial Office
MDPI AG
St. Alban-Anlage 66
Basel, Switzerland

This edition is a reprint of the Special Issue published online in the open access journal *Environments* (ISSN 2076-3298) from 2016–2017 (available at: http://www.mdpi.com/journal/environments/special_issues/VOC).

For citation purposes, cite each article independently as indicated on the article page online and as indicated below:

Author 1; Author 2. Article title. *Journal Name* **Year**, *Article number*, page range.

First Edition 2017

ISBN 978-3-03842-512-0 (Pbk)
ISBN 978-3-03842-513-7 (PDF)

Table of Contents

About the Special Issue Editor

Ki-Hyun Kim was at Florida State University for an M.S. (1984–1986) and at University of South Florida for a Ph.D. (1988–1992). He was a Research Associate at ORNL, USA (1992 to 1994). Then, he moved to Korea and stayed at Sang Ji University (1995 to 1998). In 1999, he joined Sejong University. In 2014, he moved to Dept. of Civil and Environmental Engineering at Hanyang University. His research areas focus on environmental analysis and air quality/human health management based on coordination polymers and other advanced materials. He was awarded a National Star Faculty in 2006. He is serving as an editorial member of journals (e.g., Environmental Research, Atmospheric Pollution Research, Sensors, and Scientific World). He has published more than 470 SCI journal articles.

Preface to "Volatile Organic Compounds in Environment"

The cycling of various volatile organic compounds occurring across major environmental media, i.e., air, water, and soil, pose threats to humans in diverse routes. These compounds are released not as a single constituent but as complicated mixtures into different environmental segments through numerous routes and result in unpredicted health risks ranging from various kinds of allergies to cancers due to synergistic or antagonistic interactions. Moreover, the development of antibiotic-resistant strains of microorganisms due to the presence of certain antibiotics based on aromatic systems in the environment may lead to serious biological issues.

The adverse effects of hazardous pollutants in the environment necessitate continuous efforts on their monitoring via precise identification and quantitation. Through such efforts, their pollution levels in each specific medium can be properly regulated with the application of effective control techniques. In this book, we organized various efforts directed toward monitoring and regulation of VOCs that were implemented with the advent of technological advances. In light of the environmental significance of their pollution, this book will help the readers acquire a better knowledge on all these technological issues so as to shed light on the regulation of their pollution levels under various environmental settings.

Ki-Hyun Kim
Special Issue Editor

environments

MDPI

Case Report

Toxic Volatile Organic Compounds (VOCs) in the Atmospheric Environment: Regulatory Aspects and Monitoring in Japan and Korea

Wen-Tien Tsai

Graduate Institute of Bioresources, National Pingtung University of Science and Technology, Pingtung 912, Taiwan; wttsai@mail.npust.edu.tw; Tel.: +886-8-770-3202; Fax: +886-8-774-0134

Academic Editor: Ki-Hyun Kim
Received: 6 July 2016; Accepted: 30 August 2016; Published: 7 September 2016

Abstract: In the past decades, hazardous air pollutants (HAPs), so-called air toxics or toxic air pollutants, have been detected in the atmospheric air at low concentration levels, causing public concern about the adverse effect of long-term exposure to HAPs on human health. Most HAPs belong to volatile organic compounds (VOCs). More seriously, most of them are known carcinogens or probably carcinogenic to humans. The objectives of this paper were to report the regulatory aspects and environmental monitoring management of toxic VOCs designated by Japan and Korea under the Air Pollution Control Act, and the Clean Air Conservation Act, respectively. It can be found that the environmental quality standards and environmental monitoring of priority VOCs (i.e., benzene, trichloroethylene, tetrachloroethylene, and dichloromethane) have been set and taken by the state and local governments of Japan since the early 2000, but not completely established in Korea. On the other hand, the significant progress in reducing the emissions of some toxic VOCs, including acrylonitrile, benzene, 1,3-butadiene, 1,2-dichloroethane, dichloromethane, chloroform, tetrachloroethylene, and trichloroethylene in Japan was also described as a case study in the brief report paper.

Keywords: hazardous air pollutant; volatile organic compound; air quality monitoring; regulatory system; human carcinogen

1. Introduction

In the past decades, many studies have been focused on the non-carcinogenic health effects (e.g., respiratory disease or irritation) of non-criteria air pollutants in the atmospheric environment, which are mostly released from a variety of anthropogenic sources such as petrochemical facilities, motor vehicles, metal processing/finishing industries, gas stations, and energy sectors [1]. However, a very high incidence of leukemia and lung/liver cancers has been occurring in the urban environment between the developed and developing countries [2], showing a large number of carcinogenic air pollutants in the ambient air, including benzene, 1,3-butadiene, formaldehyde, vinyl chloride, perchloroethylene, and polycyclic aromatic hydrocarbons (PAHs). Among them, benzene may be the most notable environmental carcinogens because it has been classified by the International Agency for Research on Cancer (IARC) as the Group 1 carcinogen (confirmed as a human carcinogen) [3]. As a consequence, there is an additional control program to address human health concerns resulting from exposure to hazardous air pollutants (HAPs) other than the criteria air pollutants with ambient air quality standards since the late 1980s. Thereafter, the Clean Air Act Amendments of 1990 in the USA changed the focus from HAPs to the industry sectors emitting specific HAPs and the use of Maximum Achievable Control Technology (MACT) [4,5]. With these regulation changes, the control of HAPs has become much more effective in the USA. By contrast, the Air Quality

Guidelines of the World Health Organization (WHO), first set in 1987, were advised by the European Commission as a starting point for deriving limit values on HAPs [6,7].

Volatile organic compounds (VOCs) are generally referred to as the highly reactive and/or toxic organics emitted by both human-made and natural sources due to their high volatility at normal atmospheric conditions [8]. As a result, the definition varies among scientific organizations and official agencies in different countries, but they are characterized as organic compounds that have a relatively high vapor pressure. It should be noted that VOCs are known, or may be reasonably anticipated to pose a threat of adverse air quality and human health effects [9]. In the urban environment, toxic VOCs (e.g., gasoline) and solvents also easily react with nitrogen oxides (NO_x) in the presence of sunlight to form ozone (O_3) under a series of photochemical reactions, inducing a photochemical smog in the troposphere. More significantly, many VOCs are on the list of HAPs because they pose a threat of adverse human health effects, including cancer and respiratory illness [10]. As a consequence, some states in the USA have requested to develop their own HAPs programs on toxic VOCs (e.g., acrylonitrile, benzene, epichlorohydrin, ethylene dibromide, ethylene oxide, formaldehyde, and vinyl chloride), leading to ambient air levels (AALs) or ambient air guidelines [11].

In recent years, HAP management in the Western countries and the USA has been reviewed [6,12,13]. However, there has not been any literature addressing toxic VOC management in Asian countries. Therefore, this paper is a brief report about the regulatory aspects and environmental monitoring management of toxic VOCs designated as HAPs by Japan and Korea. Furthermore, this brief report will focus on the benzene, chlorinated VOCs, and other carcinogenic VOCs, in line with international concern about the carcinogenic risks of the VOCs in recent years. Moreover, the emission reduction of some designated VOCs in Japan is also described as a case study in this review paper.

2. Legislation on Toxic VOCs in Japan and Korea

Toxic VOCs have been detected in the ambient air at low concentration levels. People have become concerned about the effect of long-term exposure to such pollution on human health such as cancer and tumors. In contrast to the European countries and North America, there was a delayed response to HAPs in the Asian countries like Japan and Korea. Table 1 summarizes the regulations for HAPs in Japan and Korea under the Air Pollution Control Act and the Clean Air Conservation Act, respectively.

Table 1. Summaries of regulations for hazardous air pollutants (HAPs) in Korea and Japan.

Country	Japan	Korea
Central authority	Ministry of the Environment	Ministry of Environment
Relevant law/act	Air Pollution Control Act	Clean Air Conservation Act
Definition	Any substance that is likely to harm human health if ingested continuously and that is a source of air pollution.	Air pollutants that are feared to directly or indirectly inflict any harm or injury on the health and property of humans or on the birth and breeding of animal and plants.
List of HAPs	22 [a]	35 [b]
Relevant measures	- Enterprise shall take the necessary measure to determine the status of emission and discharge into the atmosphere of HAPs. - The State shall endeavor to implement studies in collaboration with local public entities in order to determine the status of air pollution by HAPs, and shall periodically make public the results of human health risk evaluation.	- The central authority shall install measuring networks and constantly measure the level of air pollution. - Permissible emission levels, reduction facility installation and operation, leakage monitoring, and maintenance standards will be applicable to each industry according to the facility management standards.

[a] They are required to take priority action. Among them, the environmental air quality standards for five HAPs (i.e., benzene, trichloroethylene, tetrachloroethylene, dichloromethane, and dioxins) have been established.
[b] They were defined as specific hazardous air pollutants.

2.1. Japan

With the economic growth rate of over 10% during the period of 1960s, the Japanese experienced a strong energy demand until the oil crisis of 1973. In this regard, air pollution and degradation of environment were fired all around Japan during this period. This thus led to the promulgation of the Basic Law for Environmental Pollution Control in 1967. Thereafter, the Japanese government launched some legal enactments and revisions to enforce anti-pollution measures. Presently, the basic law governing air pollution from the emissions of soot, smoke, particulate, VOCs, hazardous air pollutants (see Table 1) and motor vehicle exhausts is the Air Pollution Control Act, which was initially passed in June 1968 and amended several times. Its main aim is to protect the health of citizens and the living environment from air pollution. Under the implementation of monitoring of the air pollution levels, it showed that air pollution is a serious environmental problem in Japan, particularly in urban areas and industrial zones [14]. Various carcinogenic air pollutants, such as benzene and chlorinated VOCs, have been detected in the ambient air in low concentrations. As a result, 22 substances are designated as priority HAPs for which measures should be taken with special action like emission monitoring and human health risk. Among them, there are 13 toxic VOCs as priority HAPs listed in Table 2.

Table 2. Summaries of toxic volatile organic compounds (VOCs) designated as HAPs in Korea and Japan. ● means that this VOC is desiganted as HAPs in Korea and/or Japan.

Toxic VOCs	Japan	Korea
Acetaldehyde	●	●
Acrylonitrile	●	●
Aniline		●
Benzene	●	●
1,3-Butadiene	●	●
Carbon tetrachloride		●
Chloroform	●	●
Chloromethyl methyl ether	●	
1,2-Dichloroethane	●	●
Dichloromethane	●	●
Dimethyl sulfide		●
Ethylbenzene		●
Ethylene oxide	●	●
Formaldehyde	●	●
Phenol		●
Propylene oxide		●
Styrene		●
Tetrachloroethylene	●	●
Trichloroethylene	●	●
Vinyl chloride	●	●

2.2. Korea

With rapid economic growth and urbanizations since the early 1970s, there was a great concern about the adverse effect of air pollutant on human health in Korea. The government began to set air quality standards for criteria pollutants as key policies under the authorization of the Clean Air Conservation Act. Air quality standards for sulfur dioxide (SO_2), carbon monoxide (CO)/nitrogen dioxide (NO_2)/total suspended particles (TSP)/ozone (O_3)/hydrocarbons, lead, benzene, and particulate matter ($PM_{2.5}$ and PM_{10}) were set in 1978, 1983, 1991, 1995, 2010, and 2011, respectively [15]. For example, benzene, known to be a human carcinogen (leukemia), exists as an ingredient in gasoline. Its air quality standard is set at 0.005 mg/m^3 based on annual average as shown in Table 3. It should be noted that benzene has been included into the National Emission Standards for Hazardous Air Pollutants (NESHAP) under the U.S. Clean Air Act Amendments of 1977.

In addition to criteria air pollutants, specific hazardous air pollutants and monitored hazardous air pollutants, defined in the Clean Air Conservation Act (seen in Table 1), have been of great concern in Korea because they could be carcinogenic to human and harmful to environmental quality. Herein, a "specific HAP" is a monitored HAP that may be harmful in the event of long-term consumption or exposure, even at low concentrations, and is deemed by committee evaluation to require atmospheric emission control. A "monitored HAP" is an air pollutant that may be harmful to human health or animal and plant growth and development, and is deemed by committee evaluation to require continuous measurement, monitoring, or observation. Accordingly, air quality management policies are shifting toward health-oriented risk and taking priority for public health. Currently, 35 substances were designated as specific hazardous air pollutants for special control and prevention [16]. Among them, 19 toxic VOCs are designated as specific HAPs in Korea as listed in Table 2. In order to reduce the health risk of carcinogenic VOCs from their fugitive emissions, prevention and control management standards for HAP-emitting facilities were enacted under the amendment of the Clean Air Conservation Act, which has been effective as of 1 January 2015 [15]. These facility management standards include permissible emission levels, reduction facility installation and operation, leakage monitoring, and maintenance standards [17].

Table 3. Environmental air quality standards for toxic VOCs in Japan and Korea.

HAPs	Environmental Air Quality Standards (Based on Annual Average)	
	Japan	Korea
Benzene	0.003 mg/m^3	0.005 mg/m^3
Trichloroethylene	0.2 mg/m^3	–
Tetrachloroethylene	0.2 mg/m^3	–
Dichloromethane	0.15 mg/m^3	–

– Not available.

3. Environmental Monitoring & Management

3.1. Japan

The Amendment of Air Pollution Control Act in 1996 required the environmental quality standards and environmental monitoring of priority HAPs by the state and local governments. Based on the carcinogenicity, physicochemical property, use, consumption, and monitoring data, benzene, trichloroethylene, tetrachloroethylene, and dichloromethane were first designated as HAPs for promotion of countermeasures and defined in the environmental quality standards (see Table 3). In compliance with the Air Pollution Control Act, local governments have monitored these toxic VOCs in the atmosphere in the past decade according to "the guideline for hazardous air pollutants monitoring" and the "manual for monitoring method of hazardous air pollutants" published by the Ministry of the Environment (MOE). To control the designated substances including benzene and trichloroethylene, the Ministry of Environment and the Ministry of Economy, Trade and Industry in Japan established a "Guideline for the Promotion of Voluntary Control of Hazardous Air Pollutants by Business Entities". Under this guideline, each industry group developed a nationwide and voluntary reduction plan in 2003. Results of the monitoring survey have been compiled by the MOE [18,19]. As shown in Table 4, the concentration levels of four toxic VOCs basically indicated a decreasing trend during this period. However, it can be found that their concentrations in ambient air fluctuated and even increased in recent years, which may be attributed to the strength of the emission source, the monitoring conditions, and locations.

Table 4. Environmental monitoring levels for toxic VOCs in the atmosphere over the past 10 years (2003–2012) in Japan [a].

HAPs	Environmental Air Quality Levels (Unit: µg/m^3) [b]									
	2003	2004	2005	2006	2007	2008	2009	2010	2011	2012
Benzene	1.9	1.8	1.7	1.7	1.5	1.4	1.3	1.1	1.2	1.2
Trichloroethylene	0.92	0.93	0.75	0.90	0.76	0.62	0.53	0.44	0.53	0.50
Tetrachloroethylene	0.38	0.38	0.28	0.31	0.25	0.23	0.22	0.17	0.18	0.18
Dichloromethane	2.4	2.6	2.1	2.8	2.3	2.3	1.7	1.6	1.6	1.6
Acrylonitrile	–	0.11	0.10	0.11	0.10	0.093	0.079	0.073	0.088	0.080
Chloroform	–	0.26	0.33	0.23	0.21	0.22	0.21	0.19	0.21	0.20
1,2-Dichloroethane	–	0.13	0.13	0.15	0.15	0.16	0.17	0.16	0.18	0.17
1,3-Butadiene	–	0.26	0.22	0.23	0.19	0.18	0.16	0.14	0.15	0.14

[a] Sources [16,17]. [b] Based on annual average concentration. – Not available.

Furthermore, in response to over thousands of toxic substances commonly used in Japan, the government implemented state and local public entity policies and measures under the Chapter II-4 ("Promotion of Countermeasures of Hazardous Air Pollutants") of the Air Pollution Control Act, including environmental monitoring guideline, health risk evaluation, and guideline values defined for other HAPs. Currently, there are four toxic VOCs for which guideline values are specified as a guide to reduce health risks resulting from HAPs in the atmosphere, especially in industrial zones. Table 4 also shows the monitoring results of these toxic VOCs for which the guideline values established [18,19]. However, it can be seen that the concentrations of 1,2-dichloroethane slightly increased, possibly due to its extensive use in the industrial sector.

3.2. Korea

As described above, Korea has set air quality standards for key air pollutants as policy objectives on air quality control and has been making efforts to satisfy these standards. In order to understand the actual air quality trend in national levels, about 200 air quality monitoring stations have been installed by each environmental management office and each city and province in urban areas or nearby industrial complex [17]. During the past two decades, all sewage treatment facilities in Korea emit significant VOCs (e.g., toluene, chlorinated hydrocarbons) from the liquid surface [20,21]. In order to reduce the fugitive emissions of toxic air pollutants, facility management standards for VOC-emitting facilities (including laundry shops, printing, painting facilities and gas stations) were enacted and have been effective as of 1 January 2015.

In Korea, the environmental policy for air quality management has recently shifted to a focus on pollution prevention and health risk-oriented management [15]. Accordingly, the Clean Air Conservation Act will be tentatively improved to reclassify HAPs into monitored HAPs (97 compounds) and specific HAPs (38 compounds, as compared to 35 compounds in Table 1) according to the following indices: toxicity, environmental and physicochemical properties, impact on ecosystems, atmospheric emission volume, ambient concentration level, and international regulations. These classifications are required to be designated by an evaluation committee. To identify the contamination state by specific HAPs in urban areas or nearby industrial complex, several organic toxics, including 13 types of VOCs and 7 types of PAHs [16,17], have been measured by 31 stations in the HAPs monitoring network. The monitored VOCs are acetaldehyde, acrylonitrile, benzene, 1,3-butadiene, carbon tetrachloride, chloroform, ethylbenzene, ethyl dichloride, formaldehyde, propylene oxide, styrene, tetrachloroethylene, and trichloroethylene.

4. Conclusions

HAPs, also called air toxics, represent a designated classification for harmful substances of anthropogenic emission sources that exist in measurable quantities in the atmospheric air, and are defined under the laws and acts of developed countries. In this paper, the legislation on toxic VOCs designated as HAPs by Japan and Korea were reviewed. It can be found that the environmental quality standards and environmental monitoring of priority VOCs (i.e., benzene, trichloroethylene, tetrachloroethylene, dichloromethane, acrylonitrile, chloroform, 1,2-dichloroethane, and 1,3-butadiene) have been set and taken by the state and local governments of Japan since the early 2000s, but not completely established in Korea. In line with the international concern about the carcinogenic risk, the results of the environmental monitoring of 8 designated VOCs in Japan were described and discussed as a case study. It was found that the monitoring data indicated a decreasing trend during this period. However, it can be found that their concentrations in ambient air fluctuated and have even increased in recent years.

In the past decades, cancer remains the most common cause of death in developed countries. A number of environmental factors have been implicated in the inductions of human cancer. However, environmental and occupational exposure to toxic VOCs (e.g., benzene, formaldehyde, and chlorinated hydrocarbons) and toxic metals (e.g., beryllium, cadmium, chromium, and nickel), especially those that are airborne, is indicative of the confirmed evidence of human carcinogenicity. As a result, the regulatory and voluntary actions to reduce stationary emissions and further study the relationship between human health risk and long-term exposure to their atmospheric concentrations should be performed. Furthermore, an advanced new concept, like the maximum available control technology (MACT) in the USA, will be able to lessen the emissions of fugitive HAPs from the petrochemical factories and refineries.

Conflicts of Interest: The author declares no conflict of interest.

References

1. Suh, H.H.; Bahadori, T.; Vallarino, J.; Spengler, D. Criteria air pollutants and toxic air pollutants. *Environ. Health Perspect.* **2000**, *108* (Suppl. S4), 625–636. [CrossRef] [PubMed]
2. Torniqvist, M.; Ehrenberg, L. On cancer risk estimation of urban air pollution. *Environ. Health Perspect.* **1994**, *102* (Suppl. S4), 173–182. [CrossRef]
3. International Agency for Research on Cancer. IARC Monographs on the Evaluation of Carcinogenic Risks to Humans. Available online: http://monographs.iarc.fr/ (accessed on 3 July 2016).
4. McClellan, R.O. Human health risk assessment: A historical overview and alternative paths forward. *Inhal. Toxicol.* **1999**, *11*, 477–518. [CrossRef] [PubMed]
5. U.S. Environmental Protection Agency. Initial List of Hazardous Air Pollutants with Modification. Available online: http://www.epa.gov/haps/initial-list-hazardous-air-pollutants-modifications (accessed on 24 August 2016).
6. Van Leeuwen, F.X.R. A European perspective on hazardous air pollutants. *Toxicology* **2002**, *181–182*, 355–359. [CrossRef]
7. World Health Organization Regional Office for Europe. Air Quality Guidelines for Europe. Available online: http://www.euro.who.int/__data/assets/pdf_file/0005/74732/E71922.pdf (accessed on 24 August 2016).
8. Godish, T.; Davis, W.T.; Fu, J.S. *Air Quality*, 5th ed.; CRC Press: Boca Raton, FL, USA, 2015.
9. Allin, C.W. *Encyclopedia of Environmental Issues: Atmosphere and Air Pollution*; Salem Press: Ipswich, MA, USA, 2011.
10. Stander, L.H. Regulatory aspects of air pollution control in the United States. In *Air Pollution Engineering Manual*, 2nd ed.; Davis, W.T., Ed.; John Wiley & Sons: New York, NY, USA, 2000; pp. 8–21.
11. Calabrese, E.J.; Kenyon, E.M. *Air Toxics and Risk Assessment*; Lewis Publishers: New York, NY, USA, 1991.
12. Hinwood, A.L.; Di Marco, P.N. Evaluating hazardous air pollutants in Australia. *Toxicology* **2002**, *181–182*, 361–366. [CrossRef]
13. Patrick, D.R. *Toxic Air Pollution Handbook*; Van Nostrand Reinhold: New York, NY, USA, 1994.
14. The Committee on Japan's Experience in the Battle against Air Pollution. *Japan's Experience in the Battle against Air Pollution*; The Pollution-Related Health Damage Compensation and Prevention Association: Tokyo, Japan, 1997.

15. Ministry of Environment (Korea). Major Policies. Available online: http://eng.me.go.kr/eng/web (accessed on 1 July 2016).

16. Baek, S.O.; Jeon, C.G. Current status and future directions of management of hazardous air pollutants in Korea- Focusing on ambient air monitoring issues. *J. Korean Soc. Atmos. Environ.* **2013**, *29*, 513–527. [CrossRef]

17. Kim, J.H.; Lee, J.J. Management changes of hazardous air pollutants sources and its proposed improvement in Korea. *J. Korean Soc. Atmos. Environ.* **2013**, *29*, 536–544. [CrossRef]

18. Ministry of the Environment (Japan). Environmental Statistics 2012. Available online: http://www.env.go.jp/en/ (accessed on 1 July 2016).

19. Ministry of the Environment (Japan). Environmental Statistics 2014. Available online: http://www.env.go.jp/en/ (accessed on 1 July 2016).

20. Ministry of Environment (Korea). 2014 Environmental Statistics Yearbook (Korean). Available online: http://eng.me.go.kr/eng/web/index.do?menuId=29&findDepth=1 (accessed on 1 July 2016).

21. Kang, K.H.; Dong, J.I. Hazardous air pollutants (HAPs) emission characterization of sewage treatment facilities in Korea. *J. Environ. Monit.* **2010**, *12*, 898–905. [CrossRef] [PubMed]

environments

MDPI

Article

Study on the Kinetics and Removal Formula of Methanethiol by Ethanol Absorption

Yinghe Jiang [1,2], Xuejun Lin [1], Wenhan Li [1,3], Xiaoying Liu [1,2] and Yuqi Wu [1,*]

1   School of Civil Engineering and Architecture, Wuhan University of Technology, Wuhan 430070, China;
    jyhe123@163.com (Y.J.); 13349923376@163.com (X.L.); felicity_yy201410@163.com (W.L.);
    xy2000225@sohu.com (X.L.)
2   Research Center of Water Supply and Water Pollution Control, Wuhan University of Technology,
    Wuhan 430070, China
3   Henan Civil Aviation Development and Investment Co. Ltd, Zhengzhou 450000, China
*   Correspondence: woshiyaya7@126.com

Academic Editors: Ki-Hyun Kim and Abderrahim Lakhouit
Received: 1 September 2016; Accepted: 21 October 2016; Published: 27 October 2016

Abstract: Biological filtration is widely used for deodorising in wastewater treatment plants. This technique can efficiently remove soluble odour-causing substances, but minimally affects hydrophobic odorants, such as methanethiol (MT) and dimethyl sulfide. Ethanol absorption capacity for MT (as a representative hydrophobic odorant) was studied, and the MT removal rate formula was deduced based on the principle of physical absorption. Results indicated that the MT removal rate reached 80% when the volume ratio of ethanol/water was 1:5. The phase equilibrium constant was 0.024, and the overall mass transfer coefficient was 2.55 kmol/m^2·h in the deodorisation tower that functioned as the physical absorption device. Examination results showed that the formula exhibited adaptability under changing working conditions. These findings provide a reference for engineering design and operation of a process for the removal of MT by ethanol absorption.

Keywords: ethanol; methanethiol (MT); absorption; removal formula; kinetic parameters

1. Introduction

With the development of the economy and society in developing countries, many wastewater treatment plants that were originally constructed in suburban districts gradually became surrounded by residential housing and business zones; large quantities of odour produced from wastewater treatment plants have seriously affected the surrounding environment. Therefore, controlling odour pollution from wastewater treatment plants has been one of the most pressing environmental issues in developing countries.

Odour-causing substances can be categorized into soluble odorants (such as H_2S and NH_3) and hydrophobic odorants, such as methanethiol (MT) and dimethyl sulfide [1,2]. Current work is mostly focused on the control of H_2S and NH_3, while there is little information about removal efficiency for hydrophobic odorants. One process may exhibit good removal performance for some odour substances, while having little effect on other odour substances [3]. For instance, biological filtration processes are extensively used for odour removal in sewage treatment plants [4,5], and can efficiently remove soluble odour-causing substances; however, they minimally influence hydrophobic odorants [6,7].

The methods for the removal of hydrophobic odorants can be divided into two categories: dry methods and wet methods. Dry methods include thermal oxidation, plasma, ultraviolet, microwave technology, etc., wherein intermediate products are present in the air and might be harmful to the environment. The products of the wet methods (such as absorption methods) are in solution, and can be controlled [8]. Wet scrubbers are economical and have high processing loads for various gases, and are widely used in deodorisation [9]. In the previous studies, the effects of MT removal by wet scrubbing process were compared using several types of solution, made of ethanol, sodium hydroxide, sodium hypochlorite, and lead acetate; accordingly, the solutions made of ethanol and lead acetate showed optimum treatment effects [10]. After absorbing, the ethanol solution could be used as a carbon source for some biological treatment processes. Further processing is often required before the reaction products of the lead acetate solution absorption method are discharged, so this process is neither economical nor environmentally friendly [11]. Findings indicate that ethanol is an ideal absorption solution.

In this research, MT served as the representative of hydrophobic odorant, ethanol was used as an absorbent, and the absorption effect on MT was investigated using a deodorisation tower model. The aims of this study are as follows:

(1) To explore the relationship between the ethanol concentration and the MT removal rate according to the principle of physical absorption, and establish the mathematical formula for MT removal by ethanol absorption.
(2) To determine the kinetic parameters of the MT-ethanol absorption system.

2. Equipment and Methods

2.1. Deodorisation Test Device

The test device for MT absorption by ethanol is shown in Figure 1.

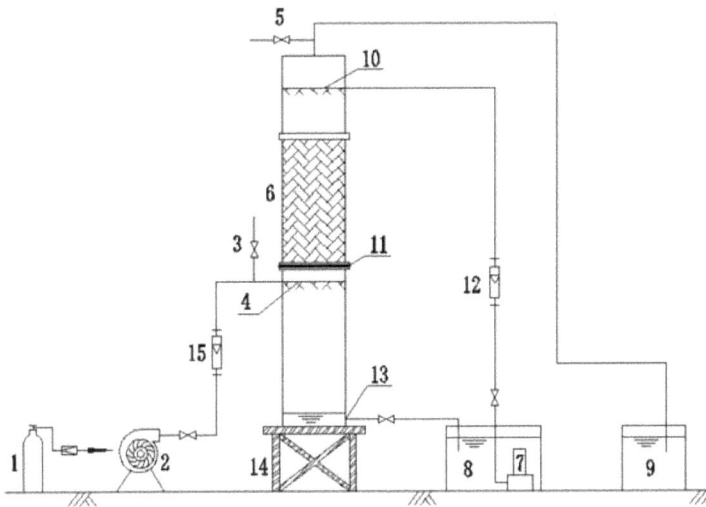

Figure 1. Schematic of the odour removal system. 1: methanethiol (MT) cylinder; 2: Blower; 3: Intake sampling port; 4: Gas distribution device; 5: Outlet sampling port; 6: Column packing; 7: Pump; 8: Absorption solution tank; 9: Outlet-gas absorbing device; 10: Sprinkler; 11: Perforated plate; 12: Liquid flowmeter; 13: Absorption solution discharge port; 14: Iron bracket; 15: Gas flowmeter.

The main component of the test unit was the counter-current ethanol absorption tower, which was made of plexiglass tubes and possesses the following dimensions: 0.1 m inner diameter; 0.8 m packing layer height; 0.5 m upper packing layer height, which could install a sprinkler system; 1 m lower packing layer height, which could install the gas distribution system and store absorption solution; and 2.3 m total height of the absorption tower. A perforated plate was set at the bottom of the packing layer, and intake pipe was placed 0.1 m below the plate. The absorption solution connection pipe was installed 5 cm above the bottom of the absorption tower to form the circulating spray system. The size of the spray liquid tank is L × B × H = 35 cm × 17cm × 33 cm (effective depth). The ceramic pall rings were used in the ethanol absorption tower; the ring exhibits high flux, low resistance, and high separation efficiency and operating flexibility. The size of the ceramic pall ring is 25 mm × 25 mm × 3 mm (diameter × height × wall thickness). The other parts are shown in Table 1.

Table 1. Types and parameters of main equipment.

Number	Equipment Name	Quantity	Type and Basic Parameters
1	Anti-corrosive type vortex pump	1	Type HG-1100 Air pressure 17.6 kPa Power 1100 w Blowing rate 180 m^3/h
2	Gas rotor flow meter	1	Type LZB-50
3	Gas rotor flow meter	1	Type LZB-3WB
4	Liquid rotor flow meter	1	Type LZB-6
5	Air sampler	1	Type QC-2A
6	MT gas tank	1	A mixed gas of MT and nitrogen (3%)

MT: Methanethiol.

2.2. Test Methods

A predetermined amount of MT in the MT cylinder was sent to the blower inlet port by the reducing valve. MT was mixed with a large quantity of air to form a preset concentration of MT odour, which entered into the tower through the gas distribution device and discharged from the top of the tower. Absorption solution was stored in the tank and sprayed from the top of the packing layer through the water pump. Gas and absorption solution liquid were in full contact in the packed bed. An air inlet pipe and an air outlet pipe in the ethanol absorption tower were provided with a sampling port to collect the original odour and the treated gas, respectively. MT removal rate was measured using MT concentration between inlet and outlet gas.

The two main factors affecting absorption efficiency were empty bed residence time (EBRT) and water–gas ratio. The minimum EBRT of the absorption method ranged between 0.4 s to 3.0 s [12]. To highlight the advantages of the absorption method, a shorter EBRT (0.6 s) was selected. Thus, the gas intake volume was 37 m^3/h. The optimal water–gas ratio was approximately 1 L/m^3, according to the preliminary study [10]. The average inlet concentration of MT was controlled below 0.1 mg/m^3, based on the monitoring data of wastewater treatment plants in Baton Rouge, Louisiana and Beijing [13,14]. Each experiment cycle was conducted according to the following steps:

(1) Absorption solution (20 L) was prepared with a preset volume ratio of ethanol/water in the absorption solution tank.
(2) The blower was operated. Mixed gas intake volume was controlled at 37 m^3/h (intake load q_G = 4700 m^3/m^2·h). MT was transmitted to the blower inlet port by controlling the reducing valve to form a preset concentration of MT odour.
(3) The spray pump was operated, and the absorption solution spraying volume was controlled at 40 L/h (spraying load q_L = 5100 L/m^2·h). The water–gas ratio was equal to 1.08.
(4) Concentrations of MT sampled from inlet and outlet pipes were measured under a sampling rate of 1.0 L/min after stable operation for 10–15 min.
(5) Waste absorption solution with a certain amount of MT absorbed in the system was discharged after each cycle.

2.3. Analysis Methods

The removal rate of MT can be calculated as follows:

$$\eta = \frac{Y_1 - Y_2}{Y_1} \tag{1}$$

where η is removal rate of MT (%); Y_1 is the content of MT in the intake port of the mixed gas (kmol MT/kmol mixed gas); and Y_2 is the content of MT in the outlet port of the mixed gas (kmol MT/kmol mixed gas).

In the absorption tower, the two following formulas could be used to calculate the packing layer height [15], in which the difference of vapour phase molar ratio is the overall mass transfer driving force:

$$Z = \frac{G_B}{K_Y \alpha} N_{OG} \tag{2}$$

$$N_{OG} = \frac{1}{1-S} ln \left[(1-S) \frac{Y_1 - mX_2}{Y_2 - mX_2} + S \right] \tag{3}$$

where:

Z: packing layer height (m);

G_B: mixed gas flow rate (kmol/m^2·h), $G_B = q_G/V_m$, where V_m is the molar volume of gas in the standard state (V_m = 22.4 L/mol);

K_Y: overall mass transfer coefficient (kmol/m^2·h);

α: effective surface area of the unit packing volume (m^2/m^3);

N_{OG}: overall gas absorption number of mass transfer units, in which the difference of the vapour phase molar ratio is the driving force;

S: desorption factor, $S = mG_B/L_S$, $L_S = xqL\rho/M$, where m is the phase equilibrium constant; L_S is the absorbent (ethanol solution) flow rate (kmol/m^2·h); x is the volume fraction of ethanol; ρ is the density of ethanol pure liquid (ρ = 0.789 kg/L); M is the molar mass of ethanol (M = 46 g/mol);

X_2: content of MT in the intake port of the ethanol (kmol MT/kmol ethanol).

Assuming:

$$k_1 = \frac{mMq_G}{\rho q_L V_m}$$

Then:

$$S = \frac{k_1}{x} \tag{4}$$

X_2 = 0 is assumed for convenience because of the low concentration of MT at the entrance of the absorption solution. Equations (1), (3), and (4) were substituted into Equation (2). Equation (5) was obtained as follows:

$$Z = \frac{q_G}{\alpha K_Y V_m} \cdot \frac{1}{1 - \frac{k_1}{x}} ln \left[(1 - \frac{k_1}{x}) \frac{1}{1 - \eta} + \frac{k_1}{x} \right] \tag{5}$$

After finishing:

$$\eta = \frac{e^{\frac{\alpha Z K_Y V_m (1 - \frac{k_1}{x})}{q_G}} - 1}{e^{\frac{\alpha Z K_Y V_m (1 - \frac{k_1}{x})}{q_G}} - \frac{k_1}{x}} \tag{6}$$

Making:

$$k_2 = \frac{\alpha Z K_Y V_m}{q_G}$$

Then:

$$\eta = \frac{e^{k_2 - \frac{k_1 k_2}{x}} - 1}{e^{k_2 - \frac{k_1 k_2}{x}} - \frac{k_1}{x}} \tag{7}$$

2.4. Monitoring Methods

The amino dimethyl aniline colorimetric method was used to measure the concentration of MT, and the detection limit was set as 0.5 μg/15 mL.

A total of 10 mL of MT absorption solution was placed in a multi-hole absorbing tube, and the sample was collected by using an atmospheric sampler. After sampling, 5 mL of absorption solution in the absorbing tube was transferred to a colorimetric tube. A total of 5 mL of fresh absorption solution was added to No. 0 absorbing tube as a comparison sample. Each colorimetric tube was added with 0.5 mL of chromogenic reagent and some distilled water up to 10 mL. The solutions in the colorimetric tubes were filtered after settling for 30 min. The absorbance of filtrate was detected at a wavelength of 500 nm.

3. Results and Discussion

3.1. Experiment Results

Absorption solution was prepared with a preset volume ratio of ethanol/water. Each ratio was subjected to three experimental cycles, and the results are shown in Table 2.

Table 2. Effect of different ethanol/water ratios on MT removal.

Volume Ratio Ethanol : Water	Cycle	Sampling Volume L	Average Inlet Concentration mg/m³	Average Outlet Concentration mg/m³	Removal Rate %
1:30	1	30	0.039	0.019	52.63
	2	60	0.037	0.019	50.00
	3	60	0.036	0.020	45.71
1:20	1	30	0.058	0.021	64.29
	2	60	0.030	0.011	62.07
	3	60	0.027	0.009	67.31
1:10	1	15	0.105	0.027	74.51
	2	30	0.058	0.014	75.00
	3	60	0.035	0.008	76.47
1:5	1	30	0.041	0.008	80.00
	2	30	0.045	0.008	81.82
	3	60	0.033	0.006	81.25
1:1	1	30	0.052	0.007	85.71
	2	60	0.042	0.007	82.93
	3	60	0.040	0.006	84.62

MT: Methanethiol.

The volume ratio of ethanol/water could be converted into the volume fraction of ethanol in the absorption solution, x. The conversion results are shown in Table 3. The effects of different volume fractions of absorption solution on the average MT removal rates, η, are shown in Figure 2.

Table 3. The results of converting ethanol/water volume ratio to volume fraction.

Volume Ratio	1:30	1:20	1:10	1:5	1:1
Volume fraction (x)	3.23%	4.76%	9.09%	16.67%	50%

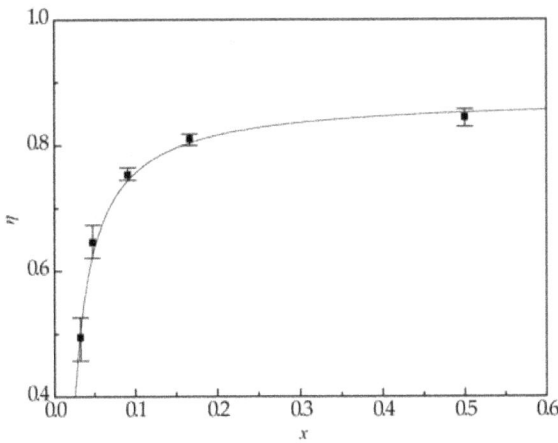

Figure 2. Effects of different volume fractions of absorption solution on methanethiol (MT) removal rate.

MT removal rate by ethanol absorption reached 80% and complied with the second class of "Emission Standards for Odour Pollutants" (GB14554-1993) when the volume fraction of the washing liquid absorption solution was 16.67% (Figure 2). In comparison with the results of Raquel Lebrero—who found that the MT removal rate decreased to 47.8% when the EBRT was 7 s in the biotrickling filter [2]—the ethanol absorption method showed great advantages. The result was also better than that of the HNO_3 solution alone, which has a removal efficiency of 73% [16].

Figure 2 indicates that the MT removal rate increased with increasing ethanol volume fraction in the absorption solution, but the increased value gradually decreased. When the concentration of ethanol solution reached 16.67%, the removal rate increased slowly. This phenomenon was mainly due to the concentration of the absorption solution no longer being a limiting factor in the absorption reaction when it reached a certain extent. Then, the removal rate hardly increased, although the absorption solution concentration continued to increase. The trend of the test results was in good agreement with the experimental data of other absorption experiments. Shen showed that the CO_2 removal rate increased with the increase of the absorption solution concentration. However, the removal rate reached a maximum when the concentration of the absorbing solution was 2 mol/L. Afterwards, the removal rate exhibited almost no increase [17]. Couvert studied the influences of hydrogen peroxide concentration, contact time, and pH on the MT removal rate, and an increase in the hydrogen peroxide concentration could evidently lead to an increase in the MT removal level, but the removal level had a limit [18].

3.2. Relationship between MT Removal Rate and Ethanol Volume Fraction

The parameters in Equation (7) were fitted by Graph software and data in Figure 2, and the results are shown in Figure 3. The fitting results were $k_1 = 0.057$ and $k_2 = 2.035$, and the relationship between the volume fraction of ethanol and the removal rate of MT is shown as Equation (8).

$$\eta = \frac{e^{2.035 - \frac{0.116}{x}} - 1}{e^{2.035 - \frac{0.116}{x}} - \frac{0.057}{x}} \tag{8}$$

$$R^2 = 0.993$$

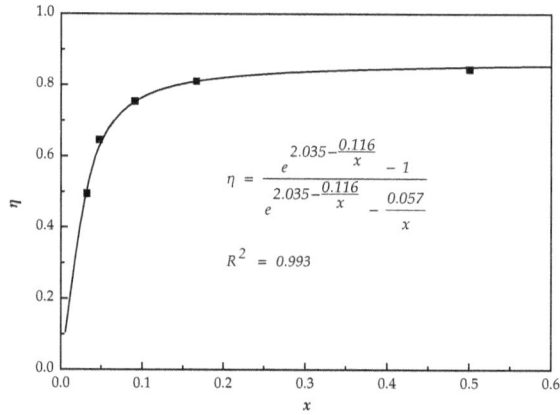

The curve figure shows:

$$\eta = \frac{e^{2.035-\frac{0.116}{x}} - 1}{e^{2.035-\frac{0.116}{x}} - \frac{0.057}{x}}$$

$$R^2 = 0.993$$

Figure 3. Curve of methanethiol (MT) removal rate under different volume fractions of ethanol solution.

3.3. Calculation of Kinetic Parameters

The set conditions in this study were $q_G = 4700 \text{ m}^3/\text{m}^2 \cdot \text{h}$, $q_L = 5100 \text{ L/m}^2 \cdot \text{h}$, $\alpha = 210 \text{ m}^2/\text{m}^3$, and $Z = 0.8$ m. Kinetic parameters m and K_Y were calculated using the value of k_1, k_2, and the following two formulas: $k_1 = \dfrac{mMq_G}{\rho q_L V_m}$ and $k_2 = \dfrac{\alpha Z K_Y V_m}{q_G}$.

Then

$$m = \frac{k_1 \rho q_L V_m}{M q_G} = 0.02363$$

making $m = 0.024$;

$$K_Y = \frac{k_2 q_G}{\alpha Z V_m} = 2.5477 \text{ kmol}/\text{m}^2 \cdot \text{h}$$

making $K_Y = 2.55 \text{ kmol}/\text{m}^2 \cdot \text{h}$.

3.4. Mathematical Formula Deduction

q_G, q_L, Z, and α were variable test parameters, and the other parameters were constants; namely, k_1 and k_2, which could be written as:

$$k_1 = \frac{mMq_G}{\rho q_L V_m} = 0.062 \frac{q_G}{q_L},$$

$$k_2 = \frac{\alpha Z K_Y V_m}{q_G} = \frac{57.07 \alpha Z}{q_G}$$

k_1 and k_2 were substituted into Equation (7). The relationship of MT removal rate and the design and operation parameters of the ethanol absorption tower could be obtained as follows:

$$\eta = \frac{a-1}{a-b} \tag{9}$$

where:

$$a = e^{\frac{57.07 \alpha Z}{q_G} - \frac{3.54 \alpha Z}{x q_L}}$$

$$b = 0.062 \frac{q_G}{x q_L}$$

3.5. Mathematical Formula Verification

3.5.1. Verification Methods

Packing type and packing layer height were maintained constant, and the air intake volume was controlled at 46 m³/h (intake load q_G = 5857 m³/m²·h), which was different from the former test. The ratio of ethanol/water and the absorption solution spraying load were changed according to the 2.2 procedure, and MT removal rate in the ethanol absorption tower was measured.

3.5.2. Verification Results

The test was conducted with a total of eight cycles, and the results are shown in Table 4.

Table 4. MT removal rate by ethanol solution absorption under different conditions.

Cycle	Volume Ratio	Intake Volume m³/h	Spraying Volume L/h	Sampling Volume L	Average Inlet Concentration mg/m³	Average Outlet Concentration mg/m³	Removal Rate %
1	1:30	46	40	10	0.072	0.041	42.86
2	1:30	46	40	20	0.036	0.021	42.86
3	1:20	46	40	10	0.062	0.031	50.00
4	1:10	46	40	20	0.036	0.010	71.43
5	1:30	46	25	10	0.082	0.062	25.00
6	1:20	46	25	10	0.051	0.031	40.00
7	1:10	46	25	10	0.072	0.030	57.14
8	1:10	46	25	10	0.082	0.031	62.50

MT: Methanethiol.

Comparing Table 4 with Table 2, the removal rate of MT decreased when the mixed gas flow increased from 37 m³/h to 46 m³/h, and when the spraying flow reduced from 40 L/h to 25 L/h, the MT removal rate decreased further. These trends agreed with the results of Shen [17]. Obviously, reducing the gas flow rate and increasing the absorbent flow rate can increase removal rate to some extent.

The ratio of ethanol to water in the absorption solution was converted into the volume fraction of ethanol in the absorption solution. Spray volume and air intake volume were converted into spray load and air intake load. Comparison of the tested removal rate and theoretical removal rate is shown in Table 5.

Table 5. Comparison between tested and theoretical removal rates.

Cycle	Volume Fraction of Ethanol %	Intake Load m³/m²·h	Spraying Load L/m²·h	Experimental Removal Rate %	Theoretical Removal Rate %	Relative Error
1	3.23	5857	5093	42.86	41.64	2.85%
2	3.23	5857	5093	42.86	41.64	2.85%
3	4.76	5857	5093	50.00	52.82	−5.64%
4	9.09	5857	5093	71.43	66.25	7.25%
5	3.23	5857	3183	25.00	27.99	−11.96%
6	4.76	5857	3183	40.00	39.14	2.15%
7	9.09	5857	3183	57.14	57.23	−0.16%
8	9.09	5857	3183	62.50	57.23	8.43%

Table 5 shows the relative errors between the theoretical removal rate calculated by Equation (9) and the tested removal rate, which are usually less than 10%. There is reason to think that under the condition of changing inlet load of mixed gas, absorption solution spraying load, packing layer height, ethanol solution ratio in absorption solution, and packing material type, Equation (9) can accurately predict MT removal rate by ethanol absorption.

3.6. Expectation

This research provided an economical and environmentally friendly method for MT removal, and it can be expanded and applied in the following aspects:

(i)   Engineering applications are needed to verify and revise the removal formula.

(ii) Waste absorption solution can be supplied as a carbon source for the odour treatment process. For example, the two-stage biological treatment-ethanol absorption method can be used to remove the mixed odours, and ethanol in the second stage can act as a carbon source for first stage biological treatment.

(iii) Waste absorption solution can be supplied as a carbon source for the sewage treatment process. Ethanol has been widely used as a supplement to biological denitrification [19], and the combination of ethanol absorption and biological denitrification will achieve maximum benefits.

4. Conclusions

This study was conducted by using a deodorisation device to achieve MT absorption with ethanol solution, providing insights into the characteristics of ethanol absorption capacity for MT. MT removal formula was deduced according to the principle of physical absorption, which could accurately calculate the MT removal rate (η) by volume fraction of ethanol (x), intake gas load (q_G), absorption solution spraying load (q_L), the height of packing layer (Z), and the effective surface area of unit packing volume (α).

The major conclusions are as follows:

(1)   When the mixed gas flow was 37 m^3/h (intake gas load q_G = 4700 m^3/m^2·h), the flow of absorption solution spraying was 40 L/h (spraying load q_L = 5100 L/m^2·h) and the ratios of ethanol/water were 1:1, 1:5, 1:10, 1:20, and 1:30. MT removal rate increased with increasing rate of ethanol volume fraction in the absorption solution. The MT removal rate reached 80% when the ratio of ethanol/water was 1:5.

(2)   In the deodorisation device of MT absorption by ethanol solution, the phase equilibrium constant m was 0.024, and the overall mass transfer coefficient K_Y was 2.55 kmol/m^2·h for engineering.

For absorption of MT in the ethanol solution, the MT removal rate formula could be calculated as:

$$\eta = \frac{a-1}{a-b}$$

$$a = e^{\frac{11984Z}{q_G} - \frac{743Z}{xq_L}}$$

$$b = 0.062\frac{q_G}{xq_L}$$

and this formula could accurately predict MT removal rate through absorption by ethanol.

Acknowledgments: The research was supported by National Nature Science Foundation of China (Grant No. 21407114). The authors deeply appreciate their financial support.

Author Contributions: Yinghe Jiang, Xuejun Lin and Wenhan Li conceived and designed the experiments; Wenhan Li performed the experiments; Yinghe Jiang and Xuejun Lin analyzed the data; Yinghe Jiang and Xiaoying Liu contributed reagents, materials and analysis tools; Xuejun Lin and Yuqi Wu wrote the paper.

Conflicts of Interest: The authors declare no conflict of interest.

References

1.   Tang, X.D. Source Identification and Sensory Quantitative Assessment of Malodorous Volatile Organic Compounds Emitted from Municipal Sewage Treatment Plant. Master's Thesis, Jinan University, Guangzhou, China, June 2011. (In Chinese)

2.   Lebrero, R.; Rodríguez, E.; Estrada, J.M.; García-Encina, P.A.; Muñoz, R. Odor abatement in biotrickling filters: Effect of the EBRT on methyl mercaptan and hydrophobic VOCs removal. *Bioresour. Technol.* **2012**, *109*, 38–45. [CrossRef] [PubMed]

3. Ong, H.T.; Witherspoon, J.R.; Daigger, G.; Quigley, C.; Easter, C.; Burrowes, P.; Sloan, A.; Adams, G.; Hargreaves, R.; Corsi, R.; et al. Odor control technologies at potws and related industrial and agricultural facilities. *Proc. Water Environ. Feder.* **2001**, *15*, 422–441. [CrossRef]

4. David, G.; Deshusses, M.A. Performance of a full-scale biotrickling filter treating H₂S at a gas contact time of 1.6 to 2.2 s. *Environ. Prog.* **2003**, *22*, 111–118.

5. Giri, B.S.; Mudliar, S.N.; Deshmukh, S.C. Treatment of waste gas containing low concentration of dimethyl sulphide (DMS) in a bench-scale biofilter. *Bioresour. Technol.* **2009**, *101*, 2185–2190. [CrossRef] [PubMed]

6. Hort, C.; Gracy, S.; Platel, V.; Moynault, L. Evaluation of sewage sludge and yard waste compost as a biofilter media for the removal of ammonia and volatile organic sulfur compounds (VOSCs). *Chem. Eng. J.* **2009**, *152*, 44–53. [CrossRef]

7. Jiang, Y.H.; Li, Q.B.; Zhou, Y.E.; Zhang, S.H.; Sang, W.J. Study on biological filter tower for removing odor from sludge dewatering room. *China Water Wastewater* **2008**, *24*, 75–77. (In Chinese)

8. Yang, S.; Wang, L.; Feng, L.; Zhao, L.; Huo, M. Wet scrubbing process for methyl mercaptan odor treatment with peroxides: Comparion of hydrogen peroxide, persulfate, and peroxymonosulfate. *Environ. Chem.* **2014**, *24*, 75–77. (In Chinese)

9. Stanley, W.B.M.; Muller, C.O. Choosing an odor control technology—Effectiveness and cost considerations. *Proc. Water Environ. Feder.* **2001**. [CrossRef]

10. Li, W.H. Experimental Research to Remove Methyl Mercaptan with Chemical Absorption Method. Master's Thesis, Wuhan University of Technology, Wuhan, China, May 2014. (In Chinese)

11. Liu, T.; Li, X.; Li, F. Development of a photocatalytic wet scrubbing process for gaseous odor treatment. *Ind. Eng. Chem. Res.* **2010**, *49*, 3617–3622. [CrossRef]

12. Ogink, N.W.M.; Melse, R.W. Air scrubbing techniques for ammonia and odor reduction at livestock operations: Review of on-farm research in the Netherlands. *Trans. ASAE* **2005**, *48*, 2303–2313.

13. Devai, I.; Delaune, R.D. Emission of reduced malodorous sulfur gases from wastewater treatment plants. *Water Environ. Res.* **1999**, *71*, 203–208. [CrossRef]

14. Huang, L.H.; Liu, J.W.; Xia, X.F.; Xu, Y.P.; Zhou, X. Research of emission characteristics of gaseous pollutants in municipal wastewater treatment plant. *Sci. Technol. Eng.* **2015**, *15*, 295–299. (In Chinese)

15. Yuan, H.X. *Separation Process and Equipment*, 1st ed.; Chemical Industry Press: Beijing, China, 2008; pp. 277–281.

16. Muthuraman, G.; Sang, J.C.; Moon, I.S. The combined removal of methyl mercaptan and hydrogen sulfide via an electro-reactor process using a low concentration of continuously regenerable Ag(II) active catalyst. *J. Hazard. Mater.* **2011**, *193*, 257–263. [CrossRef] [PubMed]

17. Shen, J.Y. Study on CO₂ Removal via Membrane Absorption Using PTFE Hollow Fiber Membranes. Master's Thesis, Zhejiang University of Technology, Zhejiang, China, January 2015. (In Chinese)

18. Couvert, A.; Charron, I.; Laplanche, A.; Renner, C.; Patria, L.; Requieme, B. Simulation and prediction of methyl-mercaptan removal by chemical scrubbing with hydrogen peroxide. *Chem. Eng. Technol.* **2006**, *29*, 1455–1460. [CrossRef]

19. Fillos, J.; Ramalingam, K.; Bowden, G.; Deur, A.; Beckmann, K. Specific denitrification rates with ethanol and methanol as sources of organic carbon. *Proc. Water Environ. Feder.* **2007**, *2007*, 251–279. [CrossRef]

environments

MDPI

Article

Investigation of Air Quality beside a Municipal Landfill: The Fate of Malodour Compounds as a Model VOC

Jacek Gębicki [1,*], Tomasz Dymerski [2] and Jacek Namieśnik [2]

[1] Department of Chemical and Process Engineering, Chemical Faculty, Gdansk University of Technology, 11/12 G. Narutowicza Str., Gdańsk 80-233, Poland
[2] Department of Analytical Chemistry, Chemical Faculty, Gdansk University of Technology, 11/12 G. Narutowicza Str., Gdańsk 80-233, Poland; tomasz.dymerski@pg.gda.pl (T.D.); jacek.namiesnik@pg.gda.pl (J.N.)
*   Correspondence: jacek.gebicki@pg.gda.pl; Tel.: +48-58-347-2752

Academic Editors: Ki-Hyun Kim and Abderrahim Lakhouit
Received: 23 November 2016; Accepted: 13 January 2017; Published: 17 January 2017

Abstract: This paper presents the results of an investigation on ambient air odour quality in the vicinity of a municipal landfill. The investigations were carried out during the spring–winter and the spring seasons using two types of the electronic nose instrument. The field olfactometers were employed to determine the mean odour concentration, which was from 2.1 to 32.2 ou/m^3 depending on the measurement site and season of the year. In the case of the investigation performed with two types of the electronic nose, a classification of the ambient air samples with respect to the collection site was carried out using the k-nearest neighbours (kNN) algorithm supported with the cross-validation method. Correct classification of the ambient air samples collected during the spring–winter season was at the level from 71.9% to 87.5% and from 84.4% to 94.8% for the samples collected during the spring season depending on the electronic nose type utilized in the studies. It was also revealed that the kNN algorithm applied for classification of the samples exhibited better discrimination abilities than the algorithms of the linear discriminant analysis (LDA) and quadratic discriminant function (QDA) type. Performed seasonal investigations proved the ability of the electronic nose to discriminate the ambient air samples differing in odorants' concentration and collection site.

Keywords: electronic nose; field olfactometry; landfill; odour; VOC (Volatile Organic Compound); Principle Component Analysis (PCA); k-Nearest Neighbours (kNN)

---

1. Introduction

Volatile organic compounds, due to their physical properties such as ease of conversion into gas state and low solubility in water, often constitute by-products in numerous industrial processes and they are sources of outdoor and indoor air pollution [1]. Moreover, many compounds are characterized by unpleasant odour, which is a cause of citizens' complaints about environment quality. The progress in urbanization and municipal infrastructure in many countries contributes to a negative phenomenon connected with the fact that residential areas are too close to such municipal objects as sewage treatment plants or municipal landfills [2–5]. The volatile organic compounds emitted from these sources and characterized by unpleasant odour include mercaptanes, sulphides (disulphides), amines, carboxylic acids, aldehydes, ketones, aliphatic and aromatic hydrocarbons [6–8]. Information on the threshold levels of odour perception for selected volatile organic pollutants emitted from sewage treatment plants or municipal landfills are gathered in Table 1 [9].

Table 1. Examples of the threshold levels of odour identification of selected odorous pollutants.

Pollutants	Threshold Level of Odour Identification (ppm)	Pollutants	Threshold Level of Odour Identification (ppm)
methanol	100	butane	1200
methylamine	4.7	octane	1.7
dimethylamine	0.34	chlorobenzene	0.68
methyl ethyl ketone	10	benzene	2.7
styrene	0.047	ethyl acetate	0.87
toluene	0.33	acetaldehyde	0.0015
methanethiol	0.00007	ethylbenzene	0.17
ethanethiol	0.0000087	α-pinene	0.018
dimethyl sulphide	0.003	limonene	0.038
acetone	42	hexane	1.5

These compounds irritate nerve cells in the human nose and they are naturally associated with danger, create a feeling of discomfort and can be a reason for negative psychosomatic symptoms. This problem could be aggravated in the future with the progress in economic and industrial development. Hence, there is a need to search for suitable tools enabling the reduction of odorous compounds emission in order to decrease odour nuisance over a particular area. The attempts to reduce the emissions of the volatile organic compounds characterized by unpleasant odour become a priority for these fields of industry, which are responsible for the emissions. Correction measures include the implementation of deodorization systems in already existing plants as well as the appropriate design and location of new-built facilities [10–12].

The acquisition of the information about the concentration levels of particular odorants in ambient air is indispensable for the complex evaluation of the condition of the natural environment [13,14]. This goal is reached by the utilization of suitable tools for the measurement and control of the level of air pollutants as well as for the identification of the presence of odorous compounds. The following devices for VOCs analysis in outdoor and indoor air can be distinguished:

- indicator tubes, in which an analyte reacts with an indicator chemical of the tube packing; the reaction results in the appearance of a colour or change of a colour of some of the tube packing. The extent (length) of the coloured zone in the tube is directly proportional to the analyte concentration and sampling period.
- stationary measurement devices such as gas chromatographs, ultraviolet (UV) and infrared (IR) spectrometers (including the ones with Fourier transformation), mass spectrometers, as well as electron capture detectors, flame ionization detectors, photo-ionization detectors and thermal conductivity detectors.
- on-line analysers (portable), which include gas chromatographs, electrochemical analysers, photo-ionization analysers, IR or UV absorption analysers, colorimeters and photometers with prepared paper tape, which changes colour upon contact with the analyte [15,16].

Two basic approaches—analytical and sensory—can be identified regarding the general classification of the techniques used for the evaluation of malodorous VOCs. The sensory techniques include the most frequently applied dynamic and field olfactometry, whereas the analytical techniques engulf gas chromatography with olfactometric detection, gas chromatography coupled with a mass spectrometer and chemical sensors matrixes (often termed electronic nose instruments).

Particular odorous substances, when present in a gas mixture, can mutually attenuate or amplify odour intensity and change the hedonic quality of odour. That is why odour impact evaluation calls for holistic analysis without quantitative identification of particular components of the odorous mixture. The analytical techniques, which fulfil this condition include [17–22]:

- olfactometric techniques,
- electronic nose technique.

The first technique consists of the utilization of the human nose as a sensor for air quality evaluation with respect to odour. Appropriately a selected group of assessors, characterized by defined odour perception, describes the odour concentration of the odorous mixtures via the determination of the dilution degree of the odorous mixture with pure air (or inert gas). This concentration is expressed in ou/m^3 units. The olfactometric techniques (dynamic olfactometry, field olfactometry) are the most frequently used ones for air evaluation with respect to odour in the countries, which possess legal regulations defining the admissible levels of odour concentration over a particular area.

The second technique relies on the detection of the odorous compounds using a set of selective/partially selective/non-selective chemical sensors. Abilities of the electronic nose are significantly limited as compared to its "biological counterpart", just to mention a necessity of suitable training and application of frequently complicated mathematical-statistical algorithms responsible for the correct interpretation of results. There are three main commercial approaches to the structure of the electronic nose:

(a) the first type, where the measurement system is comprised of the chemical sensors of one type only,
(b) the second type, where the measurement system consists of the chemical sensors of different types,
(c) the third type, where the measurement system employs chromatographic detectors and appropriately selected chromatographic columns differing in polarity of a stationary phase.

The chromatographic columns separate volatile components, which are then identified by standard chromatographic detectors. In the case of the electronic nose technique, these are short columns for ultra-fast gas chromatography that are employed, which results in a short time of analysis equal to 1–2 min [23–25]. Obtained chromatograms are analysed using data analysis methods; in this case, the chromatographic peaks play the role of the sensors.

Due to their discrimination abilities, the electronic nose instruments can be employed to discriminate the gas samples differing in quality and odour. They have found many practical applications in such fields as safety, environmental pollution, medicine, work safety regulations, the food industry, the chemical industry and other [26–47].

The authors of this paper want to compare the discrimination abilities of the electronic nose instruments built according to the second and the third concept as far as the discrimination and classification of the ambient air samples collected in the vicinity of the municipal landfill are concerned. A prototype of the electronic nose—comprised of four commercial semiconductor sensors by FIGARO Engineering Inc. (Osaka, Japan) (TGS 832, TGS 2600, TGS 2602, TGS 2603), one photoionization sensor by Ion Science Ltd (Cambridge, UK) (PPB MiniPID) and two electrochemical sensors by FIGARO Engineering Inc. (Osaka, Japan) (FECS44, FECS50)—was compared with a commercial electronic nose based on the fast gas chromatography (Fast GC)—HERACLES II. Additionally, the Nasal Ranger field olfactometers were used to assess odour concentration in ambient air at the sampling sites where the samples for electronic nose measurements had been collected.

2. Experimental

2.1. Measurement Set-Up

The electronic nose prototype was designed and involved a set of four commercial semiconductor sensors by FIGARO Engineering Inc. (TGS 832, TGS 2600, TGS 2602, TGS 2603), one photoionization sensor of PID-type (PPB MiniPID by Ion Science Ltd) and two electrochemical sensors by FIGARO Engineering Inc. (FECS44, FECS50). The measurement set-up utilized in the investigation consisted of a Tedlar bag (SKC Inc., Valley, California, CA, USA) of 5 L volume, a Tecfluid flow meter, the prototype of the electronic nose (Figure 1), a suction pump and a personal class computer. The volumetric flow rate of the air sucked from the Tedlar bag was constant and equalled 1 L/min. An analogue-to-digital converter was used to process the output signal from the sensor set of the prototype. The output signal was converted into digital form in the range from 0 to 16 bits. Measurement data were collected and archived. The values of a particular sensor signal taken for data analysis originated from the range where the sensor signal attained a steady value. The operation mode of the electronic nose was as follows: 30 s—suction of a sample, measurement; 5 min—washing of the sensors chamber with clean air.

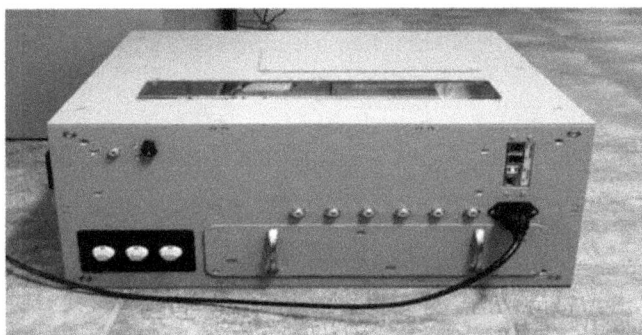

Figure 1. Prototype of the electronic nose.

The commercial electronic nose of Fast GC-type—HERACLES II (Figure 2) was built from two independent chromatographic-detection systems. The main components of these systems were two chromatographic columns characterized by different polarity of a stationary phase and two detectors of Flame Ionization Detector (FID-type). The measurement set-up consisted of the HERACLES II device, the Tedlar bag of 5 dm^3 volume and a 5 cm^3 syringe. The measurement procedure consisted of sampling the air directly from the Tedlar bag using the syringe. Then, a 5 cm^3 air sample was supplied to a proportioner. Sorption of the sample occurred behind the proportioner, inside a sorption trap of Tenax. The analytes were released from the trap after it had been heated to 270 °C and the stream was directed to two independent chromatographic-detection systems. A single analysis lasted about 100 s. The surface area of the chromatographic peaks was utilized in the analysis. In the case of the electronic nose of Fast GC-type, its operation mode followed the pattern 100 s—injection of a sample, analysis; 500 s—cleaning of the chromatographic columns with hydrogen + air at the volume ratio 1:5, respectively.

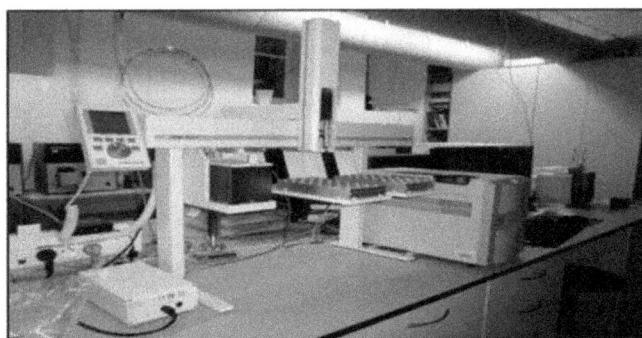

Figure 2. Electronic nose of Fast Gas Chromatography (Fast GC-type)—HERACLES II.

The principle of the operation of the device and method of testing is also described in the article [48]. Figure 2 presents the electronic nose of Fast GC-type.

Table 2 presents a comparison of both types of electronic nose instrument with respect to the operation parameters such as mass, dimensions, price, portability, gases utilized for correct operation and time of correct operation.

Table 2. Examples of the threshold levels of odour identification of selected odorous pollutants.

Parameters	Electronic Nose Prototype	Fast GC
mass (kg)	20	35
dimensions (cm × cm × cm)	30 × 20 × 50	100 × 50 × 50
portability	possible	no
operation gases utilized	clean air	hydrogen + clean air
price (euro)	ca. 8000	ca. 160,000
time of correct operation (years)	2–3	10

Four persons took part in the investigation carried out with the field olfactometers Nasal Ranger (St. Croix Sensory, Stillwater, MN, USA). These persons (a team of panellists) were selected from a larger group and trained following a standard procedure elaborated by the St. Croix Sensory, Inc. (St. Croix Sensory 2006). Moreover, the panelists were trained with respect to sensory measurements using the Nasal Ranger field olfactometers. Assessment of odorants' concentration in ambient air involved the determination of D/T values (dilution to sensing threshold), at which the odour was sensed. In order to compare the correct operation of the electronic nose prototype, Fast GG and Nasal Ranger instruments, they were tested in laboratory conditions using a reference mixture of n-butanol + air, where the concentration of n-butanol in air was 0.8 ppm v/v, which corresponded to the odour concentration at the level of 20 ou/m^3. Figure 3 schematically presents the experimental procedure of the measurement techniques—electronic nose prototype, commercial electronic nose of Fast GC-type and Nasal Ranger field olfactometer—applied for the determination of air quality beside the municipal landfill.

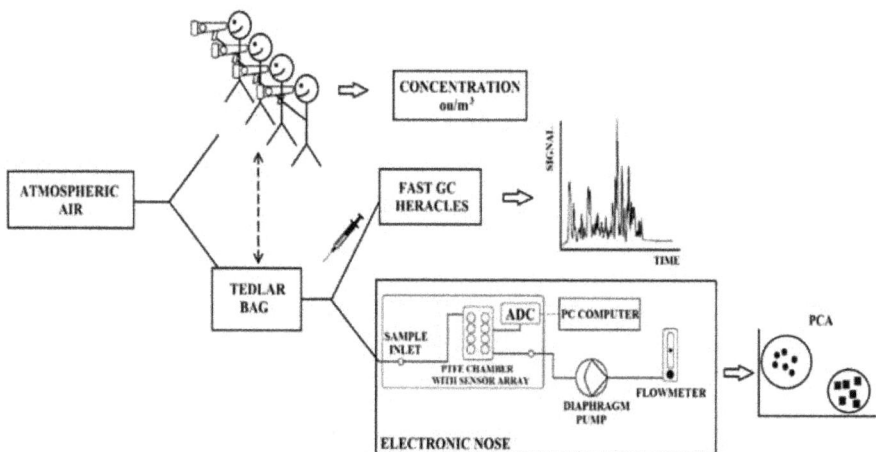

Figure 3. Schematic plan of investigation of ambient air beside the municipal landfill.

2.2. Methodology of Investigation

Investigation of air quality with respect to odorants' concentration, carried out with the electronic noses, was performed for the air samples collected around the municipal landfill in the vicinity of the

Tricity Agglomeration. The samples were collected at four control points located within 1-kilometre distance from the landfill. Localization and distribution of the air sampling points around the landfill is illustrated in Figure 4. The samples were collected during two seasons of the year: spring–winter (January–March) and spring (April–June). There was no atmospheric precipitation during the sampling operation. The samples were collected into the Tedlar bags (SKC Inc., Valley, California, CA, USA) of 5 dm^3 volume using a self-designed device called a Lung sampler. A total of 96 ambient air samples were collected around the landfill and analyzed. Analysis of the data obtained with the electronic nose prototype was performed employing free R software, part of Free Software Foundation (Free Software Foundation, Boston, MA, USA). A method of classification of the collected samples with respect to localization of the collection point involved the k-nearest neighbours algorithm (where k equaled 3); discrimination of the samples utilized the principle component analysis (PCA). Air quality investigation with respect to odorants' concentration performed with the field olfactometers was carried out at the same time and at the same control points, where the air was sampled into the Tedlar bags. A total of 384 measurements were performed with the Nasal Ranger field olfactometers.

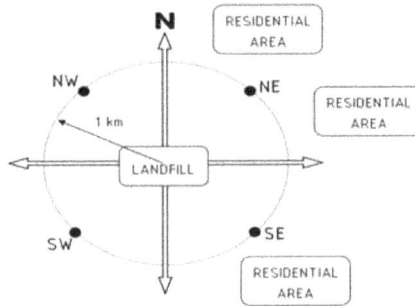

Figure 4. Map of the municipal landfill with the points of atmospheric air samples collection.

3. Results and Discussion

Figure 5 presents the PCA results for the ambient air samples collected in the vicinity of the municipal landfill when the measurements were performed with the electronic nose prototype during the spring–winter season. This time, the two-dimensional plane reveals three characteristic clusters of points, which correspond to the air samples' collection sites. One cluster is associated with the samples collected along the NE direction, the second cluster corresponds to the SE direction and the third one describes the remaining two directions, namely NW and SW. Table 3 shows an error matrix with the results of kNN (where k = 3) classification supported with the cross-validation method for the ambient air samples collected around the municipal landfill during the spring–winter season. The results of the performed classification originated from the investigation carried out with the electronic nose prototype. A total of 71.9% of the samples collected from the selected directions was correctly classified. The biggest number of correctly classified samples was 24 and they originated from the SE direction. In the remaining cases, the number of correctly classified samples was as follows: 23 for the NE direction, 13 for the NW direction and nine for the SW direction. Correctness of classification was at the level of 95.8% for the NE direction, 100% for the SE direction, 54.2% for the NW direction and 37.5% for the SW direction.

Figure 6 presents the PCA results for the ambient air samples collected in the vicinity of the municipal landfill when the measurements were performed with the commercial electronic nose of Fast GC-type during the spring–winter season. Also, in this case, there are three characteristic clusters of points on the two-dimensional plane. However, as compared to Figure 4, the clusters NE and SE are clearly resolved from the remaining points. Table 4 shows an error matrix with the results of the kNN (where k = 3) classification supported with the cross-validation method for the ambient air samples collected around the municipal landfill during the spring–winter season. The results of the performed classification originated from the investigation carried out with the commercial electronic nose of Fast GC-type. A total of 87.5% of the samples collected from the selected directions was correctly classified. The biggest number of correctly classified samples was 24 and they originated from the NE direction. In the remaining cases, the number of correctly classified samples was as follows: 23 for the SE direction, 21 for the NW direction and 16 for the SW direction. Correctness of classification was at the level of 100% for the NE direction, 95.8% for the SE direction, 87.5% for the NW direction and 66.7% for the SW direction.

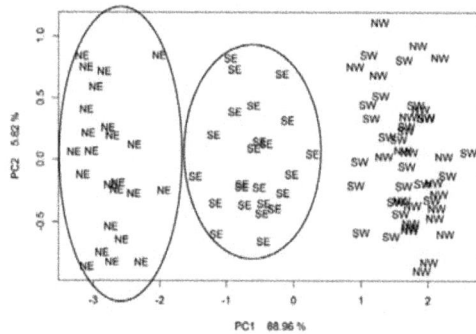

Figure 5. PCA result for the ambient air samples collected from four directions localized in the vicinity of the municipal landfill. Measurements were carried out with the electronic nose prototype during the spring–winter season.

Table 3. Cross-validation supported the k-nearest neighbours (k = 3) classification of the ambient air samples collected in the vicinity of the municipal landfill. Measurement data for classification were obtained with the electronic nose prototype during the spring–winter season.

Direction	NE	SE	SW	NW
NE	23	0	0	0
SE	1	24	1	0
SW	0	0	9	11
NW	0	0	14	13

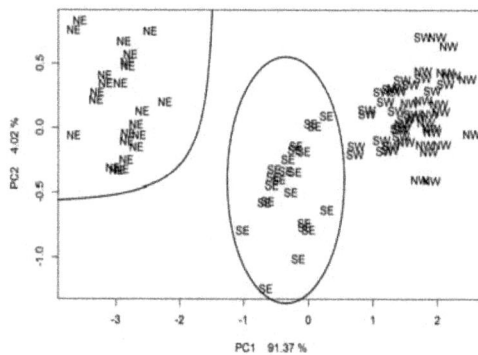

Figure 6. PCA result for the ambient air samples collected from four directions localized in the vicinity of the municipal landfill. Measurements were carried out with the commercial electronic nose of Fast GC-type during the spring–winter season.

Table 4. Cross-validation supported the kNN (k = 3) classification of the ambient air samples collected in the vicinity of the municipal landfill. Measurement data for classification were obtained with the commercial electronic nose of Fast GC-type during the spring–winter season.

Direction	NE	SE	SW	NW
NE	24	0	0	0
SE	0	23	1	0
SW	0	1	16	3
NW	0	0	7	21

Table 5 gathers the values of odour concentration C_{od} (ou/m^3) calculated as a geometric mean of the n-element set of all individual odour concentrations for a given measurement point. It can be observed that the highest values of odour concentration were estimated along the NE and SE directions. These values were 22.4 and 14.5 (ou/m^3), respectively. Concentration values determined for the remaining measurement points were similar and amounted to 2.4 (NW) and 2.1 (SW). High odour concentrations in the measurement points located along the NE and SE directions could result from many factors, including wind direction. During the investigation, predominant wind directions were north-east and south-east, which moved air masses from the area of the landfill towards the NE and SE directions where higher odorants concentrations were noticed as compared to the other measurement points.

Table 5. Values of odour concentration C_{od} (ou/m^3) calculated as a geometric mean of the n-element set of all individual odour concentrations for a particular measurement point during the spring–winter season.

Direction	NE	SE	SW	NW
concentration (ou/m^3)	22.4	14.5	2.1	2.4

Figure 7 presents the PCA results for the ambient air samples collected in the vicinity of the municipal landfill when the measurements were performed with the electronic nose prototype during the spring season. Similar to Figure 5, there are three characteristic clusters of points corresponding to the air sampling sites. One cluster represents the samples collected along the NE direction, the second cluster corresponds to the SE direction and the third cluster is associated with the remaining two directions, which are NW and SW.

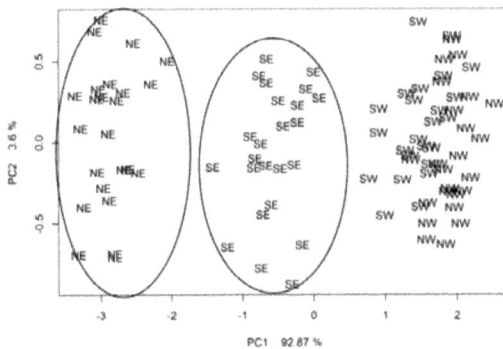

Figure 7. PCA result for the ambient air samples collected from four directions localized in the vicinity of the municipal landfill. Measurements were carried out with the electronic nose prototype during the spring season.

Table 6 shows an error matrix with the results of the kNN (where k = 3) classification supported with the cross-validation method for the ambient air samples collected around the municipal landfill during the spring season. The results of the performed classification originated from the investigation carried out with the electronic nose prototype. A total of 84.4% of the samples collected from the selected directions was correctly classified. The biggest number of correctly classified samples was 24 and they originated from the NE direction. In the remaining cases, the number of correctly classified samples was as follows: 24 for the SE direction, 18 for the NW direction and 15 for the SW direction. Correctness of classification was at the level of 100% for the NE direction, 100% for the SE direction, 75.0% for the NW direction and 62.5% for the SW direction.

Table 6. Cross-validation supported the kNN (k = 3) classification of the ambient air samples collected in the vicinity of the municipal landfill. Measurement data for classification were obtained with the electronic nose prototype during the spring season.

Direction	NE	SE	SW	NW
NE	24	0	0	0
SE	0	24	1	0
SW	0	0	15	6
NW	0	0	8	18

Figure 8 presents the PCA results for the ambient air samples collected in the vicinity of the municipal landfill when the measurements were performed with the commercial electronic nose of Fast GC-type during the spring season. However, as compared to Figure 5, the clusters NE and SE are clearly resolved from the remaining points. Table 7 shows an error matrix with the results of the kNN (where k = 3) classification supported with the cross-validation method for the ambient air samples collected around the municipal landfill during the spring season. The results of the performed classification originated from the investigation carried out with the commercial electronic nose of Fast GC-type. A total of 94.8% of the samples collected from the selected directions was correctly classified. The biggest number of correctly classified samples was 24 and they originated from the NE and SE direction. In the remaining cases, the number of correctly classified samples was as follows: 22 for the SW direction, 21 for the NW direction. Correctness of classification was at the level of 100% for the NE and SE direction, 91.7% for the SW direction, 87.5% for the NW direction.

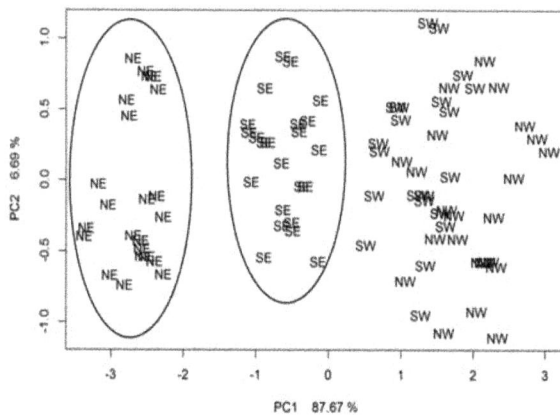

Figure 8. PCA result for the ambient air samples collected from four directions localized in the vicinity of the municipal landfill. Measurements were carried out with the commercial electronic nose of Fast GC-type during the spring season.

Table 7. Cross-validation supported the kNN (k = 3) classification of the ambient air samples collected in the vicinity of the municipal landfill. Measurement data for classification were obtained with the commercial electronic nose of Fast GC-type during the spring season.

Direction	NE	SE	SW	NW
NE	24	0	0	0
SE	0	24	0	0
SW	0	0	22	3
NW	0	0	2	21

A comparison of Tables 3, 4, 6, 7 shows that the correctness of classification of the ambient air samples collected from different control points around the municipal landfill depends on the season of the year and the measurement device enabling discrimination of the samples. The dependence between the season of the year and the correctness of classification of the ambient air samples polluted with odours has already been observed in the paper [49] where the measurements had been conducted with a commercial electronic nose of Fast GC type. It was noticed that the correctness of classification of the samples collected from various control points during the summer season was higher than in the case of the samples collected at the same points but during the winter season. Higher temperature and humidity during the summer season allows the emission of the compounds produced in anaerobic conditions inside waste dumps of municipal landfills. Table 8 gathers the values of odour concentration C_{od} (ou/m^3) calculated as a geometric mean of the n-element set of all individual odour concentrations for a given measurement point during the spring season. It can be observed that the highest values of odour concentration were estimated along the NE and SE directions. These values were 32.2 and 17.3 (ou/m^3), respectively. Concentration values determined for the remaining measurement points were similar and amounted to 2.3 (NW) and 2.2 (SW). High values of odour concentration in the control points during the spring season as compared to the spring–winter season confirm that the dominant factors causing the emission of unpleasant odorous compounds are temperature and air humidity, contributing to enhanced anaerobic processes occurring in waste dumps of municipal landfills. Earlier investigations, performed with gas chromatography allowing the identification of the compounds responsible for elevated odour concentrations, revealed the presence of the following groups of compounds: sulphides, aldehydes, ketones, amines, aliphatic and aromatic hydrocarbons, organic acids, terpenes [49]. All these compounds at the concentration levels above the odour identification threshold can undergo different phenomena of odour interaction, for instance synergism—odour intensification. Higher concentrations of odorous compounds during the spring season may arise not only from the aforementioned anaerobic or climatic processes (air temperature, relative humidity) but also due to additional compounds generated by enhanced sun radiation during this season of the year, which are characterized by high volatility and susceptibility to chemical conversions. These factors and predominant north-east and south-east winds (ca. 60% during a year) result in a higher concentration of odorous compounds along the NE and SE directions beside the municipal landfill as compared to SW and NW directions.

Table 8. Values of odour concentration C_{od} (ou/m^3) calculated as a geometric mean of the n-element set of all individual odour concentrations for a particular measurement point during the spring season.

Direction	NE	SE	SW	NW
concentration (ou/m^3)	32.2	17.3	2.2	2.3

In order to compare the discrimination abilities of the kNN algorithm (where k = 3) applied in the investigations performed with both types of the electronic nose, the obtained results of the correct classification were compared with the classification results for the kNN algorithm (where k = 5 and k = 7), the LDA (linear discriminant analysis) algorithm and the QDA (quadratic discriminant function) algorithm. The results are presented in Table 9. It can be noticed that the highest level of correct classification is exhibited by the kNN algorithm (where k = 3). The k-nearest neighbours algorithm classifies an investigated object into a particular group based on k-nearest located observations from a training set. A number of the nearest located observations (k) taken into account is pre-assumed and is an odd number. This algorithm belongs to the so-called *lazy learning algorithms* due to the fact that it investigates only a small part of the training set. It is one of the simplest classification algorithms because an unknown object is assigned to a particular group by the majority of its neighbours, the object is identified as a member of the group, which is the most popular among the object's k-nearest neighbours.

Table 9. Comparison of the discrimination abilities of the algorithms: kNN (where k = 3, k = 5, k = 7), linear discriminant analysis (LDA) and quadratic discriminant function (QDA).

Direction	Electronic Nose Prototype		Fast GC	
algorithms	spring–winter	spring	spring–winter	spring
kNN (k = 3)	71.9	84.4	87.5	94.8
kNN (k = 5)	68.7	78.5	81.2	92.0
kNN (k = 7)	66.8	76.7	80.4	90.2
LDA	70.2	81.2	85.2	93.2
QDA	71.1	82.5	86.2	93.6

4. Conclusions

Classification of the ambient air samples collected in the vicinity of the municipal landfill performed with the kNN algorithm (where k = 3) revealed that the biggest number of correctly classified samples originated from the NE and SE control points both during the spring–winter and the spring season. Correct classification of the ambient air samples for these control points was at the level of 71.9% and 84.4%, respectively, for the electronic nose prototype and at the level of 87.5% and 94.8%, respectively, for the electronic nose of Fast GC-type. Field olfactometry measurements also indicated that these control points exhibited higher odour concentration than the other measurement points. The measured values were equal to 22.4 (ou/m^3) for the NE direction and 14.5 (ou/m^3) for the SE direction during the spring–winter season and 32.2 (ou/m^3) for the NE direction and 17.3 (ou/m^3) for the SE direction during the spring season.

High values of odour concentration and higher values of the correct classification of the ambient air samples collected at different control points during the spring season as compared to the spring–winter season confirm that the main factor responsible for such a situation is climatic conditions. They include ambient air temperature, air humidity, wind direction, wind velocity and sun irradiation. A higher level of the correct classification of the air samples with the commercial electronic nose of Fast GC-type resulted from the fact that more information had been taken for analysis (data were collected from the first 18 chromatographic peaks from both columns). If each chromatographic peak (surface area) is treated as a signal from a single sensor, then the advantage of the electronic nose comprised of chromatographic columns will be obvious as far as detection abilities are concerned. However, a comparison of the operation parameters of both types of electronic nose instrument shows that the electronic nose prototype exhibits a satisfactory level of correct results in relation to its unit price. The application of the semiconductor, electrochemical and PID-type sensors improved its detection abilities and the prototype became competitive to the Fast GC-type device.

The information obtained from these season investigations will be used by the authors during further research within the frame of the project aimed at on-line monitoring of ambient air in the vicinity of the odorous compounds' emitters.

Acknowledgments: The investigations were financially supported by the Grant No. PBSII/B9/24/2013 from the National Centre for Research and Development.

Author Contributions: Jacek Gębicki developed the concept of the manuscript, Jacek Gębicki and Tomasz Dymerski conducted a study and discussion of the results, Jacek Namieśnik made substantive consultations.

Conflicts of Interest: The authors declare no conflict of interest.

References

1. Gebicki, J. Application of electrochemical sensors and sensor matrixes for measurement of odorous chemical compounds. *Trac-Trends Anal. Chem.* **2016**, *77*, 1–13. [CrossRef]
2. Capelli, L.; Sironi, S.; del Rosso, R.; Guillot, J.M. Measuring odours in the environment vs. dispersion modelling: A review. *Atmos. Environ.* **2013**, *79*, 731–743. [CrossRef]
3. Kampa, M.; Castanas, E. Human health effects of air pollution. *Environ. Pollut.* **2008**, *151*, 362–367. [CrossRef] [PubMed]
4. Sówka, I.; Sobczyński, P.; Miller, U. Impact of Seasonal Variation of Odour Emission from Passive Area Sources on Odour Impact Range of Selected WWTP. *Rocz. Ochr. Srodowiska* **2015**, *17*, 1339–1349.
5. Chang, S.T.; Chou, M.S.; Chang, H.Y. Elimination of Odors Emitted from Hot-Melting of Recycle PS by Oxidative-Reductive Scrubbing. *Aerosol Air Qual. Res.* **2014**, *14*, 293–300.
6. Gostelow, P.; Parsons, S.A.; Stuetz, R.M. Odour measurements for sewage treatment works. *Water Res.* **2001**, *35*, 579–597. [CrossRef]
7. Guillot, J.M. Odour Measurement: Focus on Main Remaining Limits Due to Sampling. *Chem. Eng. Trans.* **2012**, *30*, 295–300.
8. Capelli, L.; Sironi, S.; Barczak, R.; Grande, M.; del Rosso, R. Validation of a method for odor sampling on solid area sources. *Water Sci. Technol.* **2012**, *66*, 1607–1613. [CrossRef] [PubMed]
9. Wenjing, L.; Zhenhan, D.; Dong, L.; Jimenez, L.M.C.; Yanjun, L.; Hanwen, G.; Hongtao, W. Characterization of odor emission on the working face of landfill and establishing of odorous compounds index. *Waste Manag.* **2015**, *42*, 74–81. [CrossRef] [PubMed]
10. Capelli, L.; Sironi, S.; Del Rosso, R.; Centola, P. Predicting odour emissions from wastewater treatment plants by means of odour emission factors. *Water Res.* **2009**, *43*, 1977–1985. [CrossRef] [PubMed]
11. Naddeo, V.; Zarra, T.; Giuliani, S.; Belgiorno, V. Odour Impact Assessment in Industrial Areas. *Chem. Eng. Trans.* **2012**, *30*, 85–90.
12. Alfonsín, C.; Lebrero, R.; Estrada, J.M.; Munoz, R.; Kraakman, N.J.R.; Feijoo, G.; Moreira, M.T. Selection of odour removal technologies in wastewater treatment plants: A guideline based on Life Cycle Assessment. *J. Environ. Manag.* **2015**, *149*, 77–84. [CrossRef] [PubMed]
13. Henshaw, P.; Nicell, J.; Sikdar, A. Parameters for the assessment of odour impacts on communities. *Atmos. Environ.* **2006**, *40*, 1016–1029. [CrossRef]
14. Gebicki, J.; Byliński, H.; Namieśnik, J. Measurement techniques for assessing the olfactory impact of municipal sewage treatment plants. *Environ. Monit. Assess.* **2016**, *188*, 32. [CrossRef] [PubMed]
15. Williams, M.L. Monitoring of exposure to air pollution. *Sci. Total Environ.* **1995**, *168*, 169–174. [CrossRef]
16. Strang, C.R.; Levine, S.P.; Herget, W.F. A preliminary evaluation of the Fourier transform infrared (FTIR) spectrometer as a quantitative air monitor for semiconductor manufacturing process emissions. *Am. Ind. Hyg. Assoc. J.* **1989**, *50*, 70–77. [CrossRef]
17. Capelli, L.; Sironi, S.; Centola, P.; del Rosso, R.; Grande, M. Electronic noses for the continuous monitoring of odours from a wastewater treatment plant at specific receptors: Focus on training methods. *Sens. Actuator B-Chem.* **2008**, *131*, 53–62. [CrossRef]
18. Capelli, L.; Sironi, S.; Del Rosso, R.; Centola, P.; Bonati, S. Improvement of olfactometric measurement accuracy and repeatability by optimization of panel selection procedures. *Water Sci. Technol.* **2010**, *61*, 1267–1278. [CrossRef] [PubMed]
19. Munoz, R.; Sivret, E.C.; Parcsi, G.; Lebrero, R.; Wang, X.; Suffet, I.H.; Stuetz, R.M. Monitoring techniques for odour abatement assessment. *Water Res.* **2010**, *44*, 5129–5149. [CrossRef] [PubMed]
20. Trincavelli, M.; Coradeschi, S.; Loutfi, A. Odour classification system for continuous monitoring applications. *Sens. Actuator B-Chem.* **2009**, *139*, 265–273. [CrossRef]
21. Wilson, A.D. Diverse applications of electronic-nose technologies in agriculture and forestry. *Sensors* **2013**, *13*, 2295–2348. [CrossRef] [PubMed]

22. Boeker, P. On "Electronic Nose" methodology. *Sens. Actuator B-Chem.* **2014**, *204*, 2–17. [CrossRef]
23. Delgado-Rodríguez, M.; Ruiz-Montoya, M.; Giraldez, I.; López, R.; Madejón, E.; Díaz, M.J. Use of electronic nose and GC-MS in detection and monitoring some VOC. *Atmos. Environ.* **2012**, *51*, 278–285. [CrossRef]
24. Cheng, H.; Qin, Z.H.; Guo, X.F.; Hu, X.S.; Wu, J.H. Geographical origin identification of propolis using GC-MS and electronic nose combined with principal component analysis. *Food Res. Int.* **2013**, *51*, 813–822. [CrossRef]
25. Xiao, Z.; Yu, D.; Niu, Y.; Chen, F.; Song, S.; Zhu, J.; Zhu, G. Characterization of aroma compounds of Chinese famous liquors by gas chromatography-mass spectrometry and flash GC electronic-nose. *J. Chromatogr. B* **2014**, *945*, 92–100. [CrossRef] [PubMed]
26. Haddi, Z.; Amari, A.; Alami, H.; El Bari, N.; Llobet, E.; Bouchikhi, B. A portable electronic nose system for the identification of cannabis-based drugs. *Sens. Actuator B-Chem.* **2011**, *155*, 456–463. [CrossRef]
27. Alizadeh, T.; Zeynali, S. Electronic nose based on the polymer coated SAW sensors array for the warfare agent simulants classification. *Sens. Actuator B-Chem.* **2008**, *129*, 412–423. [CrossRef]
28. Brudzewski, K.; Osowski, S.; Pawłowski, W. Metal oxide sensor arrays for detection of explosives at sub-parts-per million concentration levels by the differential electronic nose. *Sens. Actuator B-Chem.* **2012**, *161*, 528–533. [CrossRef]
29. Stuetz, R.M.; Fenner, R.A.; Engin, G. Characterisation of wastewater using an electronic nose. *Water Res.* **1999**, *33*, 442–452. [CrossRef]
30. Bourgeois, W.; Stuetz, R.M. Measuring wastewater quality using a sensor array prospects for real-time monitoring. *Water Sci. Technol.* **2000**, *41*, 107–112.
31. Deshmukh, S.; Bandyopadhyay, R.; Bhattacharyya, N.; Pandey, R.A.; Jana, A. Application of electronic nose for industrial odors and gaseous emissions measurement and monitoring—An overview. *Talanta* **2015**, *144*, 329–340. [CrossRef] [PubMed]
32. Fend, R.; Bessant, C.; Williams, A.J.; Woodman, A.C. Monitoring haemodialysis using electronic nose and chemometrics. *Biosens. Bioelectron.* **2004**, *19*, 1581–1590. [CrossRef] [PubMed]
33. Bernabei, M.; Pennazza, G.; Santonico, M.; Corsi, C.; Roscioni, C.; Paolesse, R.; Di Natale, C.; D'Amico, A. A preliminary study on the possibility to diagnose urinary tract cancers by an electronic nose. *Sens. Actuator B-Chem.* **2008**, *131*, 1–4. [CrossRef]
34. D'Amico, A.; Pennazza, G.; Santonico, M.; Martinelli, E.; Roscioni, C.; Galluccio, G.; Paolesse, R.; Di Natale, C. An investigation on electronic nose diagnosis of lung cancer. *Lung Cancer* **2010**, *68*, 170–176. [CrossRef] [PubMed]
35. Dang, L.; Tian, F.; Zhang, L.; Kadri, C.; Yin, X.; Peng, X.; Liu, S. A novel classifier ensemble for recognition of multiple indoor air contaminants by an electronic nose. *Sens. Actuator B-Chem.* **2014**, *207*, 67–74. [CrossRef]
36. Cynkar, W.; Cozzolino, D.; Dambergs, B.; Janik, L.; Gishen, M. Feasibility study on the use of a head space mass spectrometry electronic nose (MS e-nose) to monitor red wine spoilage induced by *Brettanomyces* yeast. *Sens. Actuator B-Chem.* **2007**, *124*, 167–171. [CrossRef]
37. El Barbri, N.; Llobet, E.; El Bari, N.; Correig, X.; Bouchikhi, B. Electronic Nose Based on Metal Oxide Semiconductor Sensors as an Alternative Technique for the Spoilage Classification of Red Meat. *Sensors* **2008**, *8*, 142–156. [CrossRef] [PubMed]
38. Dymerski, T.; Gebicki, J.; Wardencki, W.; Namiesnik, J. Application of an Electronic Nose Instrument to Fast Classification of Polish Honey Types. *Sensors* **2014**, *14*, 10709–10724. [CrossRef] [PubMed]
39. Rajamaki, T.; Arnold, M.; Venelampi, O.; Vikman, M.; Rasanen, J.; Itavaara, M. An Electronic Nose and Indicator Volatiles for Monitoring of the Composting Process. *Water Air Soil Pollut.* **2005**, *1*, 71–87. [CrossRef]
40. Rudnitskaya, A.; Legin, A. Sensor systems, electronic tongues and electronic noses, for the monitoring of biotechnological processes. *J. Ind. Microbiol. Biotechnol.* **2008**, *35*, 443–451. [CrossRef] [PubMed]
41. Li, J.; Feng, H.; Liu, W.; Gao, Y.; Hui, G. Design of A Portable Electronic Nose system and Application in *K* Value Prediction for Large Yellow Croaker (*Pseudosciaena crocea*). *Food Anal. Methods* **2016**, *9*, 2943–2951. [CrossRef]
42. Zhuang, L.; Guo, T.; Cao, D.; Ling, L.; Su, K.; Hu, N.; Wang, P. Detection and classification of natural odors with an in vivo bioelectronic nose. *Biosens. Bioelectron.* **2015**, *67*, 694–699. [CrossRef] [PubMed]
43. Zheng, L.; Zhang, J.; Yu, Y.; Zhao, G.; Hui, G. Spinyhead croaker (*Collichthys lucidus*) quality determination using multi-walled carbon nanotubes gas-ionization sensor array. *J. Food Meas. Charact.* **2016**, *10*, 247–252. [CrossRef]

44. Lu, Y.; Li, H.; Zhuang, S.; Zhang, D.; Zhang, Q.; Zhou, J.; Dong, S.; Liu, Q.; Wang, P. Olfactory biosensor using odorant-binding proteins from honeybee: Ligands of floral odors and pheromones detection by electrochemical impedance. *Sens. Actuator B-Chem.* **2014**, *193*, 420–427. [CrossRef]

45. Jiang, J.; Li, J.; Zheng, F.; Lin, H.; Hui, G. Rapid freshness analysis of mantis shrimps (*Oratosquilla oratoria*) by using electronic nose. *J. Food Meas. Charact.* **2016**, *10*, 48–55. [CrossRef]

46. Lin, H.; Jiang, J.; Zheng, F.; Hui, G. Hairtail (*Trichiurus haumela*) freshness determination method based on electronic nose. *J. Food Meas. Charact.* **2015**, *9*, 541–549.

47. Zheng, L.; Gao, Y.; Zhang, J.; Li, J.; Yu, Y.; Hui, G. Chinese Quince (*Cydonia oblonga* Miller) Freshness Rapid Determination Method Using Surface Acoustic Wave Resonator Combined with Electronic Nose. *Int. J. Food Prop.* **2016**, *19*, 2623–2634. [CrossRef]

48. Dymerski, T.; Gębicki, J.; Namieśnik, J. Comparison of Evaluation of Air Odour Quality in Vicinity of Petroleum Plant Using a Prototype of Electronic Nose Instrument and Fast GC Technique. *Chem. Eng. Trans.* **2016**, *54*, 259–264.

49. Gębicki, J.; Dymerski, T.; Namieśnik, J. Application of Ultrafast Gas Chromatography to recognize odor nuisance. *Environ. Prot. Eng.* **2016**, *42*, 97–106.

environments

MDPI

Article

Photocatalytic Degradation of Toluene, Butyl Acetate and Limonene under UV and Visible Light with Titanium Dioxide-Graphene Oxide as Photocatalyst

Birte Mull, Lennart Möhlmann and Olaf Wilke *

Bundesanstalt für Materialforschung und-prüfung (BAM), Berlin 12205, Germany; birte.mull@bam.de (B.M.); Lennart.Moehlmann@gmx.de (L.M.)
* Correspondence: olaf.wilke@bam.de; Tel.: +49-30-8104-1422

Academic Editors: Ki-Hyun Kim and Abderrahim Lakhouit
Received: 21 November 2016; Accepted: 13 January 2017; Published: 25 January 2017

Abstract: Photocatalysis is a promising technique to reduce volatile organic compounds indoors. Titanium dioxide (TiO_2) is a frequently-used UV active photocatalyst. Because of the lack of UV light indoors, TiO_2 has to be modified to get its working range shifted into the visible light spectrum. In this study, the photocatalytic degradation of toluene, butyl acetate and limonene was investigated under UV LED light and blue LED light in emission test chambers with catalysts either made of pure TiO_2 or TiO_2 modified with graphene oxide (GO). TiO_2 coated with different GO amounts (0.75%–14%) were investigated to find an optimum ratio for the photocatalytic degradation of VOC in real indoor air concentrations. Most experiments were performed at a relative humidity of 0% in 20 L emission test chambers. Experiments at 40% relative humidity were done in a 1 m^3 emission test chamber to determine potential byproducts. Degradation under UV LED light could be achieved for all three compounds with almost all tested catalyst samples up to more than 95%. Limonene had the highest degradation of the three selected volatile organic compounds under blue LED light with all investigated catalyst samples.

Keywords: photocatalysis; emission test chamber; volatile organic compounds

---

1. Introduction

People from the Western world spend most of their time indoors, either in houses or in transportation, hence, the indoor air quality is of particular importance [1]. Though, volatile organic compounds (VOC) are ubiquitous in the indoor air, emitting, e.g., from building materials, wall and floor coverings and interior equipment. Those emissions can have a negative impact on the indoor air quality, health and wellbeing of occupants. This effect is described as sick-building-syndrome [2]. An approach to improve the indoor air quality, which has come more and more into focus during the last years, is the photocatalytic degradation of pollutants by photoredox catalysis. Photocatalytic active coatings, mostly made of the semiconductor titanium dioxide (TiO_2) [3], have been developed and tested for indoor air use. TiO_2 in its anatase modification has a band gap of 3.2 eV and can hence be activated under UV-light (λ = 387 nm). By light irradiation with a corresponding wavelength, electrons from the valence band are transferred to the conduction band and as a result electron-hole pairs are formed (Figure 1).

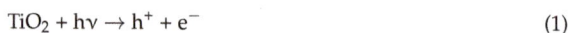

$$TiO_2 + h\nu \rightarrow h^+ + e^- \tag{1}$$

The excited electrons can proceed in a single electron reduction and, in the presence of O_2, form a superoxide radical anion $O_2^{\bullet-}$. Simultaneously, the electron holes (h^+) can react with H_2O to yield hydroxyl radicals $OH\bullet$ (single electron oxidation). These resulting radicals are highly reactive and can mineralize organic substrates to CO_2 and H_2O as generalized in equations 2 and 3 [3,4].

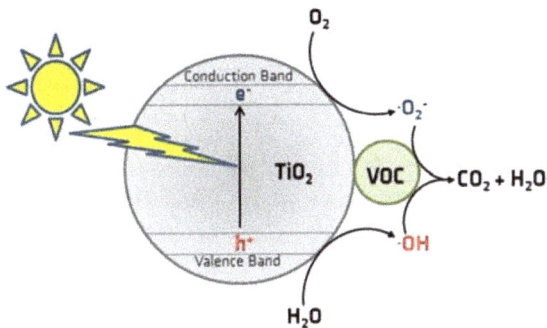

Figure 1. Photocatalytic degradation mechanism.

$$OH\bullet + VOC + O_2 \rightarrow nCO_2 + mH_2O \quad (2)$$

$$O_2^{\bullet-} + pollutant \rightarrow\rightarrow\rightarrow CO_2 + H_2O \quad (3)$$

Since TiO_2 can only be activated under UV light it has to be doped or modified to shift its working range in the visible light spectrum. Banerjee et al. published an article summarizing and describing possibilities, e.g., metal and non-metal doping, dye sensitization, combination of TiO_2 with reduced graphene oxide, to improve the visible light activity of TiO_2 [5]. Graphene is a relatively new but promising material [6] and its oxide in combination with TiO_2 is already known to successfully catalyze degradation of VOC, methyl orange and methylene blue under visible light [7–9]. With its high surface area of 2600 $m^2 \cdot g^{-1}$ [10] graphene sheets are highly reactive thus enhancing the photocatalytic activity of TiO_2 due to the fact that more π–π interactions are possible [8]. The band gap has to be below 3.0 eV to get absorption under visible light [11], the light absorption is extended by increasing the amount of GO [9]. However, if the amount of graphene in the catalyst system is too high the photocatalytic activity is reduced [12,13].

One additional substantial aspect for the reactivity is the relative humidity in the emission test chamber which can both increase or decrease the photocatalytic degradation of VOC. According to Zhao and Yang [14] and Mo et al. [15] too little water vapor results in a retard of the degradation; whereas too much water vapor is also not appropriate for the degradation process. In the second case, the water molecules and the VOC compete for adsorption positions on the TiO_2 surface. Furthermore, the effect is also depended on the inlet concentration of the VOC [16]. However, the optimum relative humidity level seems to be different depending on the investigated compound and used catalyst [17].

This work focuses on the photocatalytic degradation and investigation of potential byproducts of toluene, butyl acetate and limonene on pure TiO_2 and TiO_2 modified with different amounts of graphene oxide (GO; 0.75%, 1%, 2.5%, 5%, 10%, 14%) under UV and visible (blue) LED light. These compounds were selected because of their frequent occurrence in the indoor environment [2]. Experiments were performed in emission test chambers of various volumes under controlled climatic conditions. A reference VOC atmosphere in the range of real indoor air concentrations was created by using a gas mixing system (GMS) [18].

2. Materials and Methods

2.1. Chemicals

Toluene (108-88-3, for organic residue analysis) and methanol (67-56-1, ultra resi-analyzed) were purchased from J. T. Baker, butyl acetate (123-86-4, 99+%) and (R)-(+)-limonene (5989-27-5, 97%) from Aldrich and naphthalene d_8 in methanol (2000 $\mu g \cdot mL^{-1}$) from Restek. A graphene oxide water dispersion was purchased from Graphenea (4 mg·mL^{-1}, monolayer content >95%). The TiO$_2$ was commercially available N-doped anatase.

2.2. Analytical Parameters and Instrumentation

Quantification was done using calibration curves (8 points) in combination with an internal standard (naphthalene d$_8$ in methanol). Each tube was loaded with 1 µL of internal standard and methanol was flushed off with a nitrogen stream for 10 min (100 mL·min^{-1}) before air samples were taken.

Air sampling was done periodically with Tenax® TA sorption tubes (Gerstel; 1 L with 50 or 100 mL·min^{-1}). For the determination of lower volatile byproducts air sampling was done with Carbograph® 1TD (Markes International, Llantrisant, UK), a multi-bed tube containing Tenax® TA, Carbograph® 1TD and Carboxen® 1000, DNPH-cartridges (Supelco, Bellefonte, PA, USA; 2,4-dinitrophenylhydrazine) and silica cartridges (self-made with silica purchased from Supelco; 30 L and 60 L with 1 L·min^{-1}).

Tenax TA® tubes were measured with a TD-GC-FID system (thermal desorption-gas chromatograph-flame ionization detector) for a quantitative determination. To determine the formation of byproducts measurements were done with a TD-GC-MS (mass spectrometer) system, HPLC (high performance liquid chromatography) and IC (ion chromatography) for a quantitative determination. The limit of quantification for VOC is 1 µg·m^{-3}, for formaldehyde 5 µg·m^{-3}, for acetaldehyde 7 µg·m^{-3}, for formic acid 4 µg·m^{-3} and for acetic acid 5 µg·m^{-3}.

Tenax TA® measurements were carried out on a GC-FID system from Agilent Technologies (Santa Clara, CA, USA) (GC: 6890 N) equipped with a thermal desorption system (TDS) from Gerstel temperature program: 40 °C for 1 min, 40 °C·min^{-1} to 280 °C for 5 min) and a cooled injection system (CIS 4; temperature program: −100 °C for 0.01 min, 12 °C·s^{-1} to 280 °C for 3 min). Chromatographic separation of the VOC was done on a Rtx-Volatiles column (Restek; 30 m × 320 µm × 1.5 µm) with the following temperature program: splitless, 60 °C for 1 min, 10 °C·min^{-1} to 180 °C for 3 min, 20 °C·min^{-1} to 260 °C for 5 min (run time 25 min).

Carbograph® 1TD and multi-bed tubes were thermally desorbed in a thermal desorption system (TD-100™, Markes International) and trapped afterwards in a cold trap made of glass. This was connected to a gas chromatograph 6890 N (Agilent Technologies) coupled with a 5975B mass spectrometer (Agilent Technologies) for quantification and identification. The gas chromatographic separation was done on an Rxi-5MS column (Restek, 60 m × 0.25 mm × 0.25 µm) with a constant helium flow (Alphagaz Air Liquide) of 1.5 mL·min^{-1}. Thermal desorption was done at 300 °C for 10 min followed by 320 °C for 5 min with a flow of 50 mL·min^{-1}. The cold trap started at −5 °C and was heated up to 320 °C for 15 min. The transfer line was set at 180 °C. A split (4.3:1) injection on the GC column was done and the GC oven program started at 50 °C (initial time 0.5 min) and heated up with 6 °C·min^{-1} to 300 °C (holding time 5 min). The MS (transfer line at 300 °C, EI-source at 230 °C and quadrupole at 150 °C) operated in SCAN-mode (m/z 35–450) without solvent delay.

HPLC measurements were done on 1100 Series (Agilent Technologies) equipped with an ULTRASEP ES ALD column (125 mm × 2 mm) with a pre-column (10 mm × 2 mm). The column temperature is set to 25 °C. Acetonitrile (solvent A) and a mixture of water and tetrahydrofuran (1:16.7; solvent B) were the used solvents. The pump program was the following: 30% solvent A, 70% solvent B with a flow of 0.5 mL·min^{-1}; 30% solvent A, 70% solvent B with a flow of 0.5 mL·min^{-1} for 5 min; 32% solvent A, 68% solvent B with a flow of 0.5 mL·min^{-1} for 10 min; 32% solvent A, 68% solvent B with a flow of 0.6 mL·min^{-1} for 30 min; 83% solvent A, 17% solvent B with a flow of 0.6 mL·min^{-1} for 55 min; 30% solvent A, 70% solvent B with a flow of 0.6 mL·min^{-1} for 60 min; 30% solvent A, 70% solvent B with a flow of 0.5 mL·min^{-1} for 70 min. The detector wavelength was set at 365 nm and 380 nm.

IC measurements were carried out on a Thermo Scientific ICS 2100 instrument equipped with a Dionex IONPAC® AS18 (2 × 250 mm) column. The oven was set to 30 °C. A KOH-solution was used as eluent generator with water as solvent. The chromatographic separation was done as follows: 0 min: eluent concentration: 1 mmol; 23 min: eluent concentration: from 1 mmol to 50 mmol; 33 min: eluent concentration: from 50 mmol to 1 mmol; 43 min: end run.

2.3. Chamber Parameters

VOC degradation was investigated in 20 L and 1 m^3 emission test chambers. The 20 L emission test chambers operated under the following conditions. A translucent plate of borosilicate glass was used as chamber cover and the light sources were placed on top of it. The chambers were equipped with a stirrer to guarantee a homogenous air distribution. A GMS was used to generate stable toluene, butyl acetate and limonene concentrations. The GMS consists of a mixing chamber (V = 24 L), into which the gaseous VOC are transferred by a nitrogen stream. This mixing chamber is supplied with an air flow (114 $L \cdot h^{-1}$) of purified and dry air to get a concentration reduction of the generated VOC. Two 20 L emission test chambers were connected to the mixing chamber (Figure 2). An air flow of 200 $mL \cdot min^{-1}$ was transferred into these chambers where the degradation experiments were carried out. The photocatalytic degradation was investigated at T = 23 °C, relative humidity (r.h.) = 0%, air change rate n = 0.6 h^{-1} and a loading factor of 0.5 $m^2 \cdot m^{-3}$, which describes the ratio of catalyst surface (m^2) to emission test chamber volume (m^3). These parameters were not changed during all experiments.

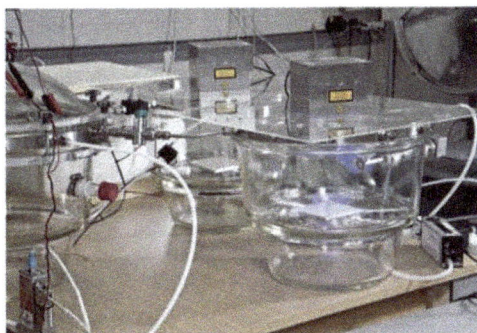

Figure 2. Experimental setup of the gas mixing chamber with two connected 20 L emission test chambers. Both chambers are equipped with ceramic tiles and one is irradiated with UV light.

The 1 m^3 emission test chamber was directly connected to the GMS without interposing of a mixing chamber. Hence, the VOC concentration was adjusted by the chambers air change rate, which was set at 1 h^{-1}. Degradation experiments were performed at a relative humidity of 40% with a loading factor of 0.03 $m^2 \cdot m^{-3}$.

2.4. Light Sources

As light sources a UV LED lamp (λ = 365 nm; irradiance of 1.9 $mW \cdot cm^{-2}$ in a distance of 25 cm) and a blue LED lamp (λ = 455 nm; irradiance of 1.9 $mW \cdot cm^{-2}$ in a distance of 25 cm) were used for the experiments in the 20 L emission test chambers. Degradation experiments in a 1 m^3 emission test chamber used an UV lamp (λ_{min} = 297 nm; irradiance of 5.6 $W \cdot m^{-2}$ in a distance of 25 cm) and blue LED lights (λ = 444 nm; irradiance of 5.9 $W \cdot m^{-2}$ in a distance of 25 cm).

2.5. Degradation Experiments under UV and Visible Light

2.5.1. General Information

Customary ceramic tiles (10 cm × 10 cm) were coated with the catalyst (GO-modified and pure TiO_2). Seven different GO-TiO_2 amounts (A–G, Table 1) were tested, each with two diverse catalyst loadings (39 and 78 mg, samples A–F). The catalysts were applied by spraying and either cured for 60 min at 100 °C (samples A–F) or for 360 min at 470 °C (sample G). For sample G, only one loading amount was tested, which differs from the others. For sample D a third loading applied by doctor blade technique was tested as well. All irradiation experiments lasted up to 7 days and air sampling was done periodically. Before each experiment the chambers were tested for blank values. Furthermore, it was tested beforehand if the VOC are degraded by photolysis under UV LED and blue LED light, which they did not. Hence, degradation was achieved photocatalytically.

2.5.2. Degradation Experiments in 20 L Emission Test Chambers

The chambers were loaded with the respective catalyst samples under exposure of the reference VOC atmosphere. Unexpectedly, after the installation of samples A to E the limonene and butyl acetate concentrations dropped significantly (down to 0 $\mu g \cdot m^{-3}$). The reason for that was a strong adsorption of these compounds onto the catalyst surface. Over time, the concentration increased again due to saturation; however, it took between 1 and 5 weeks until equilibrium and thereby a constant VOC atmosphere was reached. In contrast to that the adsorption of toluene was negligible. Since it could take up to 5 weeks until an adsorption-desorption equilibrium was achieved, sample B-1 and B-2 were pre-stored in a toluene, butyl acetate and limonene atmosphere in a desiccator for up to 24 days. After this time saturation of the catalyst samples was achieved and the degradation experiments could be started within one day after loading the catalyst samples in the emission test chambers. After the concentrations had stabilized, the degradation experiments were started by irradiation of the catalyst samples with either UV LED or blue LED light. For an overview of the starting concentrations see Table 1.

Table 1. Maximum initial concentrations.

Sample	GO (%)	Catalyst Loading (mg)	Toluene ($\mu g \cdot m^{-3}$)	Butyl Acetate ($\mu g \cdot m^{-3}$)	Limonene ($\mu g \cdot m^{-3}$)
A-1	0	39	230	130	110
B-1	0.75	39	100	140	170
B-2	0.75	78	100	220	390
C-1	1	39	190	120	60
C-2	1	78	230	140	40
D-1	2.5	39	140	130	100
D-2	2.5	78	150	110	90
D-3 *	2.5	5	90	140	80
E-1	5	39	80	120	100
E-2	5	78	90	140	40
F-1	10	39	220	190	420
F-2 [#]	10	78	230	170	330
G-1	14	10	80	120	140

* Applied by doctor blade technique; [#] investigated only under UV light.

2.5.3. Degradation Experiments in a 1 m³ Emission Test Chamber

The main focus of the experiments in a 1 m³ emission test chamber was the determination of potential byproducts. For that reason, concentrations higher than usual indoor air concentrations were generated (Table 2). The experiments were done with two samples, A-1 (pure TiO_2) and D-1 (2.5% GO), under UV LED light at 40% relative humidity. Three catalyst samples were loaded into the emission test chamber simultaneously. For the determination of potential byproducts air sampling was done additionally with Carbograph® 1TD tubes, multi-bed tubes, DNPH-cartridges and silica cartridges.

Environments **2017**, *4*, 9

Table 2. Initial concentrations.

Sample	GO (%)	Catalyst Loading (mg)	Toluene ($\mu g \cdot m^{-3}$)	Butyl Acetate ($\mu g \cdot m^{-3}$)	Limonene ($\mu g \cdot m^{-3}$)
A-1	0	39	290	380	360
D-1	2.5	39	290	390	330

3. Results

3.1. Results of the Experiments in 20 L Emission Test Chambers

A degradation of toluene, butyl acetate and limonene was achieved with pure TiO_2 and TiO_2-GO catalyst samples under UV LED light (Table 3). Within 24 h the degradation maxima were achieved and stayed constant until the light was switched off. By using GO amounts of 10% and 14% the VOC degradation was inhibited under UV and blue LED light. This phenomenon is already described in the literature by Aleksandrzak et al. [13] for the photocatalytic degradation of phenol. The photocatalytic degradation was highest when using single layered graphene, while by increasing the amount of layers, the photocatalytic degradation decreased. It is assumed that this effect is a result of surface and electronic properties of graphene oxide with TiO_2, e.g., the transport of electron-hole pairs between TiO_2 and GO is reduced. Under blue LED light limonene was degraded in most cases but butyl acetate could only be degraded by six samples, whereas toluene could not be degraded at all (see Table 3). Limonene has the highest degradation rate of the three VOC, which becomes most obvious by comparing the results obtained under blue LED light (Table 3). When using pure TiO_2 only limonene could be degraded under blue LED light, but when GO was introduced into the catalytic system butyl acetate was also degraded at least in some cases. This might be a result of the fact that graphene oxide enhances the reactivity of the catalytic system due to the formation of more π–π interactions between catalyst and VOC.

Table 3. Degradation (%) of toluene, butyl acetate and limonene under UV and blue LED light.

Sample	GO (%)	Catalyst Loading (mg)	UV LED Light			Blue LED Light		
			Toluene (%)	Butyl Acetate (%)	Limonene (%)	Toluene (%)	Butyl Acetate (%)	Limonene (%)
A-1	0	39	>95	>95	>95	0	0	92
B-1	0.75	39	>95	>95	>95	0	12	>95
B-2	0.75	78	>95	>95	>95	0	10	>95
C-1	1	39	>95	>95	>95	0	0	>95
C-2	1	78	>95	>95	>95	0	39	>95
D-1	2.5	39	>95	>95	>95	0	14	>95
D-2	2.5	78	>95	>95	>95	0	34	>95
D-3 *	2.5	5	>95	>95	>95	0	0	92
E-1	5	39	93	>95	>95	0	11	>95
E-2	5	78	94	>95	>95	0	0	>95
F-1	10	39	23	92	94	0	0	14
F-2	10	78	18	81	84	n.m.	n.m.	n.m.
G-1	14	10	0	26	>95	0	0	10

* Applied by doctor blade technique; n.m.: not measured.

3.2. Results of the Experiments in a 1 m^3 Emission Test Chamber

Degradation experiments were performed with samples A-1 and D-1. Both catalyst samples showed a slight degradation of all VOC after 24 h (Table 4). The degradation obtained in the 1 m^3 emission test chamber was lower than in the 20 L emission test chambers (Table 3), which might be caused by the lower loading factor (20 L emission test chamber: L = 0.5 $m^2 \cdot m^{-3}$; 1 m^3 emission test chamber: L = 0.03 $m^2 \cdot m^{-3}$) and the higher air change rate (20 L emission test chamber: n = 0.6 h^{-1};

1 m^3 emission test chamber: n = 1 h^{-1}) resulting in a shorter reaction time of the VOC molecules with the catalyst.

Byproducts detected for sample A-1 were formaldehyde, acetic acid, formic acid and traces of crotonaldehyde, octenal, nonenal and undecenal. Degradation of sample D-1 resulted in similar byproducts namely formaldehyde, acetaldehyde, acetic acid, formic acid and traces of octenal (Table 4). Formaldehyde, acetaldehyde, formic acid and acetic acid are already literature known byproducts of photocatalytic toluene degradation [19]. The same applies to the formed byproducts of limonene [20].

Table 4. Degradation of volatile organic compounds (VOC) in a 1 m^3 emission test chamber at 40% relative humidity under UV light.

Sample	Toluene (%)	Butyl Acetate (%)	Limonene (%)	Formaldehyde ($\mu g \cdot m^{-3}$)	Acetaldehyde ($\mu g \cdot m^{-3}$)	Formic Acid ($\mu g \cdot m^{-3}$)	Acetic Acid ($\mu g \cdot m^{-3}$)
A-1	5	10	12	9	n.d.	14	32
D-1	9	21	15	18	10	55	43

n.d.: Not detected.

3.3. Influence of Relative Humidity

According to the literature [14,15] a certain amount of relative humidity is necessary to obtain photocatalytic degradation by using TiO$_2$ as catalytic material (see Section 1). The experiments presented in this article were either performed at 0% relative humidity (20 L emission test chamber) or 40% relative humidity (1 m^3 emission test chamber). A better VOC degradation was obtained at 0% relative humidity for catalyst samples made of pure TiO$_2$ and combinations of TiO$_2$ and GO as can be seen by comparing the results in Tables 3 and 4. This outcome disproves the theory that a certain level of relative humidity is inevitable to obtain photocatalytic degradation. Consequently, in the presented experiments the superoxide radical anion might lead to photocatalytic VOC degradation. This radical is formed of the atmospheric oxygen in the air and is also strong enough to degrade VOC. This is supported by an article from Sun et al. [21] where the superoxide anion radical is described as the active species which degrades toluene photocatalytically.

4. Conclusions

Various TiO$_2$-GO composite photocatalysts have been tested for the degradation of reference VOC toluene, butyl acetate and limonene at typical indoor air concentration levels at 0% relative humidity. During irradiation experiments it has been revealed, that a modification of TiO$_2$ with trace amounts of GO (0.75%–5%) is beneficial for its photocatalytic performance under blue light. Increasing the amount of GO (10% or higher) however significantly reduces its activity both under UV as well as blue light irradiation.

Adsorption on the catalyst surface strongly affected the time for the degradation experiment. Toluene showed no adsorption onto the catalyst but butyl acetate and limonene adsorbed on its surface. It took up to 5 weeks until an adsorption-desorption equilibrium was achieved. For the investigation of photocatalytic degradation, it is very important to distinguish between adsorption and degradation. When measuring the decrease of VOC concentrations, it is necessary to ensure that the decrease is induced only by degradation.

In an experiment with VOC air concentrations of approximately 300 $\mu g \cdot m^{-3}$ at 40% relative humidity it was investigated if byproducts are formed during the degradation process. Byproducts already known from literature, e.g., formaldehyde, acetaldehyde, formic acid and acetic acid, could be detected. Unfortunately, those byproducts might be more harmful than the original VOC. For that reason, it would be helpful to develop catalysts which do not form byproducts but only water and carbon dioxide. The described test method is a useful tool for such development.

Acknowledgments: The research project was supported by the Federal Ministry of Economics and Technology under grant number KF2201064ZG3.

Author Contributions: Birte Mull and Olaf Wilke conceived and designed the experiments. Lennart Möhlmann performed the experiments and analyzed the data. Birte Mull wrote the paper.

Conflicts of Interest: The authors declare no conflict of interest.

References

1. Klepeis, N.E.; Nelson, W.C.; Ott, W.R.; Robinson, J.P.; Tsang, A.M.; Switzer, P.; Behar, J.V.; Hern, S.C.; Engelmann, W.H. The national human activity pattern survey (NHAPS): A resource for assessing exposure to environmental pollutants. *J. Expo. Anal. Environ. Epidemiol.* **2001**, *11*, 231–252. [CrossRef] [PubMed]
2. Wolkoff, P.; Clausen, P.A.; Jensen, B.; Nielsen, G.D.; Wilkins, C.K. Are we measuring the relevant indoor pollutants? *Indoor Air* **1997**, *7*, 92–106. [CrossRef]
3. Mo, J.; Zhang, Y.; Xu, Q.; Lamson, J.J.; Zhao, R. Photocatalytic purification of volatile organic compounds in indoor air: A literature review. *Atmos. Environ.* **2009**, *43*, 2229–2246. [CrossRef]
4. Pelaez, M.; Nolan, N.T.; Pillai, S.C.; Seery, M.K.; Falaras, P.; Kontos, A.G.; Dunlop, P.S.M.; Hamilton, J.W.J.; Byrne, J.A.; O'Shea, K.; et al. A review on the visible light active titanium dioxide photocatalysts for environmental applications. *Appl. Catal. B Environ.* **2012**, *125*, 331–349. [CrossRef]
5. Banerjee, S.; Pillai, S.C.; Falaras, P.; O'Shea, K.E.; Byrne, J.A.; Dionysiou, D.D. New insights into the mechanism of visible light photocatalysis. *J. Phys. Chem. Lett.* **2014**, *5*, 2543–2554. [CrossRef] [PubMed]
6. Geim, A.K.; Novoselov, K.S. The rise of graphene. *Nat. Mater.* **2007**, *6*, 183–191. [CrossRef] [PubMed]
7. Jo, W.-K.; Kang, H.-J. Titanium dioxide-graphene oxide composites with different ratios supported by pyrex tube for photocatalysis of toxic aromatic vapors. *Powder Technol.* **2013**, *250*, 115–121. [CrossRef]
8. Khalid, N.R.; Ahmed, E.; Hong, Z.L.; Sana, L.; Ahmed, M. Enhanced photocatalytic activity of graphene-TiO_2 composite under visible light irradiation. *Curr. Appl. Phys.* **2013**, *13*, 659–663. [CrossRef]
9. Nguyen-Phan, T.-D.; Pham, V.H.; Shin, E.W.; Pham, H.-D.; Kim, S.; Chung, J.S.; Kim, E.J.; Hur, S.H. The role of graphene oxide content on the adsorption-enhanced photocatalysis of titanium dioxide/graphene oxide composites. *Chem. Eng. J.* **2011**, *170*, 226–232. [CrossRef]
10. Stankovich, S.; Dikin, D.A.; Dommett, G.H.; Kohlhaas, K.M.; Zimney, E.J.; Stach, E.A.; Piner, R.D.; Nguyen, S.T.; Ruoff, R.S. Graphene-based composite materials. *Nature* **2006**, *442*, 282–286. [CrossRef] [PubMed]
11. Leary, R.; Westwood, A. Carbonaceous nanomaterials for the enhancement of TiO_2 photocatalysis. *Carbon* **2011**, *49*, 741–772. [CrossRef]
12. Zhang, Y.; Tang, Z.-R.; Fu, X.; Xu, Y.-J. TiO_2-graphene nanocomposites for gas-phase photocatalytic degradation of volatile aromatic pollutant: Is TiO_2-graphene truly different from other TiO_2-carbon composite materials? *ACS Nano* **2010**, *4*, 7303–7314. [CrossRef] [PubMed]
13. Aleksandrzak, M.; Adamski, P.; Kukulka, W.; Zielinska, B.; Mijowska, E. Effect of graphene thickness on photocatalytic activity of TiO_2-graphene nanocomposites. *Appl. Surface Sci.* **2015**, *331*, 193–199. [CrossRef]
14. Zhao, J.; Yang, X. Photocatalytic oxidation for indoor air purification: A literature review. *Build. Environ.* **2003**, *38*, 645–654. [CrossRef]
15. Mo, J.H.; Zhang, Y.P.; Xu, Q.J. Effect of water vapor on the by-products and decomposition rate of ppb-level toluene by photocatalytic oxidation. *Appl. Catal. B Environ.* **2013**, *132*, 212–218. [CrossRef]
16. Demeestere, K.; Dewulf, J.; van Langenhove, H. Heterogeneous photocatalysis as an advanced oxidation process for the abatement of chlorinated, monocyclic aromatic and sulfurous volatile organic compounds in air: State of the art. *Crit. Rev. Environ. Sci. Technol.* **2007**, *37*, 489–538. [CrossRef]
17. Korologos, C.A.; Philippopoulos, C.J.; Poulopoulos, S.G. The effect of water presence on the photocatalytic oxidation of benzene, toluene, ethylbenzene and m-xylene in the gas-phase. *Atmos. Environ.* **2011**, *45*, 7089–7095. [CrossRef]
18. Richter, M.; Jann, O.; Horn, W.; Pyza, L.; Wilke, O. System to generate stable long-term voc gas mixtures of concentrations in the ppb range for test and calibration purposes. *Gefahrst. Reinhalt. Luft* **2013**, *73*, 103–106.
19. Mo, J.; Zhang, Y.; Xu, Q.; Zhu, Y.; Lamson, J.J.; Zhao, R. Determination and risk assessment of by-products resulting from photocatalytic oxidation of toluene. *Appl. Catal. B Environ.* **2009**, *89*, 570–576. [CrossRef]

20. Ourrad, H.; Thevenet, F.; Gaudion, V.; Riffault, V. Limonene photocatalytic oxidation at ppb levels: Assessment of gas phase reaction intermediates and secondary organic aerosol heterogeneous formation. *Appl. Catal. B Environ.* **2015**, *168–169*, 183–194. [CrossRef]

21. Sun, J.J.; Li, X.Y.; Zhao, Q.D.; Ke, J.; Zhang, D.K. Novel $V_2O_5/BiVO_4/TiO_2$ nanocomposites with high visible-light-induced photocatalytic activity for the degradation of toluene. *J. Phys. Chem C* **2014**, *118*, 10113–10121. [CrossRef]

environments

MDPI

Article

Negative Reagent Ions for Real Time Detection Using SIFT-MS

David Hera [1], Vaughan S. Langford [1], Murray J. McEwan [1,2,*], Thomas I. McKellar [1] and Daniel B. Milligan [1]

[1]   Syft Technologies Ltd., 3 Craft Pl, Christchurch 8242, New Zealand; david.hera@syft.com (D.H.); vaughan.langford@syft.com (V.S.L.); thomas.mckellar@syft.com (T.I.M.); daniel.milligan@syft.com (D.B.M.)
[2]   Department of Chemistry, University of Canterbury, Christchurch 8140, New Zealand
*   Correspondence: murray.mcewan@canterbury.ac.nz; Tel.: +64-27-461-0336

Academic Editors: Ki-Hyun Kim and Abderrahim Lakhouit
Received: 17 November 2016; Accepted: 8 February 2017; Published: 15 February 2017

Abstract: Direct analysis techniques have greatly simplified analytical methods used to monitor analytes at trace levels in air samples. One of these methods, Selected Ion Flow Tube-Mass Spectrometry (SIFT-MS), has proven to be particularly effective because of its speed and ease of use. The range of analytes accessible using the SIFT-MS technique has been extended by this work as it introduces five new negatively charged reagent ions (O^-, OH^-, O_2^-, NO_2^-, and NO_3^-) from the same microwave powered ion source of moist air used to generate the reagent ions traditionally used (H_3O^+, NO^+, and O_2^+). Results are presented using a nitrogen carrier gas showing the linearity with concentration of a number of analytes not readily accessible to positive reagent ions (CO_2 from ppbv to 40,000 ppmv, sulfuryl fluoride and HCl). The range of analytes open to the SIFT-MS technique has been extended and selectivity enhanced using negative reagent ions to include CCl_3NO_2, SO_2F_2, HCN, CH_3Cl, PH_3, $C_2H_4Br_2$, HF, HCl, SO_2, SO_3, and NO_2.

Keywords: Selected Ion Flow Tube; VOCs; SIFT-MS; negative reagent ions; acid gases

1. Introduction

Most of the sensitive and specific methods for analysing dilute mixtures of volatile compounds in air utilize mass spectrometric detection. Traditionally gas chromatography-mass spectrometry (GC/MS) methods have been the method of choice for more than 50 years [1]. In GC/MS, physical separation of the volatile analytes is achieved in the capillary column of the GC. Transfer of the column output into the mass spectrometer is followed by ionization and then analysis providing identification and quantification. It sounds simple, but in practice a good knowledge of the methodology is required. For example, when mixtures of chemically different analytes are examined, different polarity chromatographic columns must be used which increases the time for analysis. On some occasions—depending on the analyte concentrations—column overload can occur. At the other extreme, when very dilute mixtures are being analysed it is often necessary to go through a pre-concentration step by first adsorbing the analyte mixture onto a substrate and subsequently heating the substrate to release the concentrated analytes at a later stage. This pre-concentration step is then followed by passage of the sample through the chromatographic column and then mass spectrometric analysis. The entire process can take longer than an hour and each step in the process can reduce the accuracy of the analysis.

In the past few years, several new and more direct analytical techniques for monitoring volatiles in air have become available [2]. The advantage of some of the direct techniques is that they may avoid the delays that occur in the more conventional but labour-intensive GC/MS and Liquid Chromatography-Mass Spectrometry (LC/MS) techniques arising from chromatography. One of these newer techniques, Selected Ion Flow Tube-Mass Spectrometry (SIFT-MS), is the topic of this review. The analytical application of SIFT-MS was pioneered by Spanel and Smith in 1996 [3]. SIFT-MS utilizes known ion-molecule reactions of mass-selected reagent ions with an analyte. The mass-selected reagent ions (traditionally H_3O^+, NO^+, and O_2^+) are introduced into a flow tube at low energy into a carrier gas where they undergo chemical reactions with the analytes in the gas sample that is drawn directly into the SIFT-MS flow tube at a known rate. The ensuing reagent ion-analyte ion-molecule reaction enables identification of the analyte in seconds and provides quantitation from the ratio of peak heights of the analyte product ion(s) relative to the reagent ion [4].

Since its introduction by Spanel and Smith, the SIFT-MS technique has been applied to monitoring volatile analytes in air utilizing the speed and ease of operation of the technique in many applications ranging from breath analysis [5–9], disease diagnosis [10–16], environmental monitoring [17–25] (including hazardous air pollutants) [26], and food and flavour characterization [27–37], to name but a few.

In this review of applications of the SIFT-MS technique, we discuss the principles on which it is based and go on to describe some of the developments that have occurred recently to make it perhaps the simplest to operate and the fastest direct analytical technique for analysing mixtures of volatile compounds in air. A comparative study between SIFT-MS and GC/MS in which 25 Volatile Organic Compounds (VOCs) from the toxic organic compendium methods (TO-14A and TO-15) from the United States Environmental Protection Agency (EPA) were jointly examined by each technique [38]. The investigation found that 85% of the VOCs examined in the test mixture were within 35% of their stated concentrations by the SIFT-MS measurements without any calibration. The concentrations of these VOCs were found simply by utilizing the known ion-molecule chemistry of the mass-selected reagent ions and the ratio of the reaction product ion counts to the reagent ion counts [4].

A side-by-side comparison of GC/MS and SIFT-MS confirmed that SIFT-MS provides a viable alternative to GC/MS down to pptv levels but with the marked additional benefit of eliminating the pre-concentration steps required by GC/MS for trace level gas analysis [38]. In addition, the linear concentration range found in SIFT-MS is up to six orders of magnitude compared to the one–two orders of magnitude linear relationship with concentrations typically found in GC/MS [38].

Up until 2015, the traditional reagent ions used in SIFT-MS were H_3O^+, NO^+, and O_2^+, which were chosen because they are readily generated from a microwave discharge of moist air and they do not react with the major constituents of air. However, one disadvantage of restricting the reagent ions to just these three positively charged ions is that not all volatile analytes react with them. There are a number of environmentally significant volatiles (such as HF, HCl, SO_2, and SO_3) that are ubiquitous pollutants in some industries but do not react with H_3O^+, NO^+, and O_2^+. Similarly, quite a number of different small molecules are used as fumigants in shipping containers. Some of these common fumigants are hydrogen cyanide, HCN; ethylene dibromide, $C_2H_4Br_2$; ethylene oxide, C_2H_4O; methyl bromide, CH_3Br; chloropicrin, CCl_3NO_2; formaldehyde, HCHO; acetaldehyde, CH_3CHO; phosphine, PH_3 and Vikane (sulfuryl fluoride), SO_2F_2. Several of these fumigants such as Vikane are unreactive with positive reagent ions. Others such as phosphine have ambiguities when other analytes such as H_2S are present with positive ion reagents. For reasons such as these, we have extended the range of reagent ions available for analysis of samples by adding five additional negative ions O^-, OH^-, O_2^-, NO_2^-, and NO_3^-. These ions can also be generated from the same microwave ion source of moist air as the positive reagent ions. We also examined the consequences of changing the bath gas from helium to nitrogen.

2. Experimental Method

A schematic outline of a commercial SIFT-MS instrument is shown in Figure 1. It has four distinct regions. The ion source region traditionally operates at 400 millitorr of moist air generating the major ions H_3O^+, NO^+, and O_2^+ which are then mass-selected by the upstream quadrupole. The mass-selected ions then pass into the flow tube where the reagent ion-analyte reactions take place. The unreacted reagent ions together with the product ions resulting from the reactions are mass-selected by the downstream quadrupole and counted.

In order to generate negative reagent ions, the pressure in the ion source was increased to above 700 millitorr and a negative voltage gradient (up to 500 V) was introduced into the microwave discharge. Adjustments to lens voltages required for changing between positively and negatively charged reagent ions can be done in milliseconds. However, changes to the ion source pressure conditions take longer (seconds). For this reason, several strategic changes were also introduced into the software controlling the moisture and pressure conditions in the ion source allowing easy switching between positive and negative ion formation. Five negatively charged reagent ions were found.

The reagent ions OH^- and O_2^- are usually generated from moist air and O^-, NO_2^-, and NO_3^- from a dry air source. These negative reagent ions are mass-selected by the upstream quadrupole in the same way as the positive reagent ions before entering the flow tube. The reagent ions can generally be interchanged within 8 ms, although switching between moist and dry negative ion sources takes several seconds.

The carrier gas is added through the inlet labelled "carrier inlet" in Figure 1, and the sample is simply drawn into the flow tube via the sample inlet at a known rate defined by the sample capillary dimensions.

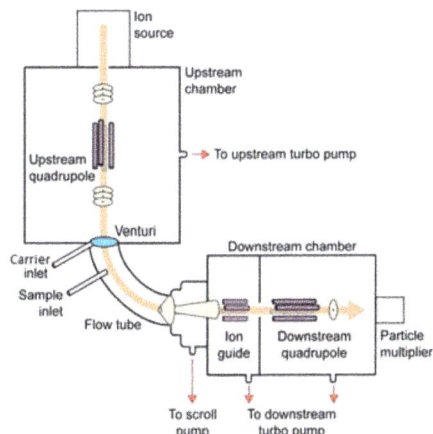

Figure 1. Schematic of a commercial Selected Ion Flow Tube-Mass Spectrometry (SIFT-MS) instrument from Syft Technologies Ltd showing the arrangement of the various components for the Voice200 model.

2.1. Nitrogen Carrier Gas

When nitrogen was used as the carrier gas, the pressure in the flow tube was usually reduced from 600 millitorr (with helium as the carrier gas) to around 450 millitorr. Entry of the reagent ions into the flow tube is assisted by means of a Venturi inlet [4,39] which facilitates the transmission of ions against the pressure gradient. Although helium had been the carrier gas of choice, in view of the unreliable supplies in some countries and its cost, we have utilized nitrogen which can be supplied by a nitrogen generator. A consequence of using nitrogen means it is necessary to know the rate coefficients of any three body reactions if relevant in the chemistry that is used to evaluate analyte concentrations. Most of the data in the literature has been evaluated for a helium carrier gas and although these data may be applied to nitrogen for binary exothermic ion molecule reactions, some adjustments may need to be made for three body reactions in nitrogen.

2.2. Negative Reagent Ions

Typical reagent ion counts for positive ions are usually greater than 10^7 cps, which enables concentrations of analytes as low as pptv to be monitored in real time [40], although the negative ions have a slightly lower abundance.

A comparison between the positive and negative mode of ion source operation is informative. For the positive reagent ions of H_3O^+, NO^+, and O_2^+, typical analogue currents after transmission through the upstream quadrupole as measured on the last lens before transmission into the flow tube are around 10.5 nA for each ion. After transmission through the flow tube, the ion current measured at the lens sampling the ions at the downstream end is reduced to ~110 pA. For the negative reagent ions of O^-, OH^-, O_2^-. and NO_2^-, the analogue currents transmitted by the upstream quadrupole are typically ~−6 nA reducing to ~−70 pA after passage along the flow tube, but with some variation depending on the reagent ion. Prior to injection of the reagent ions from the upstream quadrupole chamber, the lens voltages are optimized to minimise the presence of any ions to less than ~5% of the mass-selected reagent ion counts [41]. As noted in the next section, the reactions of some negative reagent ions with atmospheric CO_2 yield cluster ions that can participate in secondary ion chemistry. Those instruments operating with a nitrogen carrier gas instead of helium require a little more care in adjusting the lens voltages and ion source pressure than for positive reagent ions, in order to optimize the negative reagent ion signal.

In the experiments that follow, negative reagent ions were mass selected by the upstream quadrupole and passed into the flow tube where a nitrogen carrier gas carried them along to the downstream quadrupole. Here, the product and reagent ions of the reactions examined were counted. Certified concentration mixtures of specific analytes were obtained from the suppliers specified in the next section, and gas mixtures of these specified analytes were added through the sample inlet at a known flow rate into the reaction tube. The known rate coefficients of the negative reagent ions with those analytes were used to monitor their concentrations using the SIFT-MS instrument which were then compared with the concentrations of the mixtures specified by the suppliers using the methods discussed in the literature for SIFT-MS [5]. In the case of the fumigants examined in this work (with the exception of sulfuryl fluoride), rather than using gas mixtures of analytes from reference mixtures at certified concentrations, permeation tubes (supplied by Kintek, La Marque, TX, USA) were used that supply certified concentrations of the fumigant at a specified temperature and oven flowrate for the nitrogen carrier gas.

2.3. Chemicals

Two certified mixtures of CO_2 were obtained from a commercial supplier (CAC Gas and Instrumentation, Arndell Park, NSW, Australia) of 500 ppmv (in nitrogen) and 40,000 ppmv (in synthetic air) both with a stated accuracy of ±2%.

A certified mixture of sulfuryl fluoride (Vikane) (5.02 ppmv in air (±2%) was supplied by Scott Marrin, Riverside, CA, USA). Certified concentrations of the acid gases in nitrogen HCl (10 ppm ± 10%); SO_2 (10 ppm ± 10%); H_2S (25 ppm ± 5%), and NO_2 (10 ppm ± 10%) were obtained from CAC Gas, NSW, Australia. A certified mixture of HF in nitrogen (10 ppm ± 20%) was obtained from Matheson, Aendell Park, Australia and SO_3 (>99% contained in a stabilizer) was obtained from Aldrich, St. Louis, MO, USA. The remaining fumigants (chloropicrin, HCN, CH_3Cl, PH_3, and dibromoethane) were obtained from known concentrations of these fumigants in nitrogen obtained from permeation tubes operating at their calibrated flows and temperatures as supplied by Kintek (La Marque, TX, USA).

3. Results and Discussion

All the results shown here were obtained using a nitrogen carrier gas and at a flow tube temperature of 120 °C. We show the mass spectra of four of the negative reagent ions in the flow tube with and without an air sample of laboratory air flowing into the flow tube through the reaction tube sample inlet and also with a breath exhalation (Figure 2a, O^-; Figure 2b, OH^-; Figure 2c, O_2^-; and Figure 2d, NO_2^-). The reason for the inclusion of the sample air and breath exhalation is to show the presence of CO_2. CO_2 is present naturally in air, and slow termolecular association reactions of the negative reagent ions occur with it. The products of these negative ion reactions with CO_2 reactions are therefore present in every scan of a sample in air. Base levels of CO_2 in New Zealand are around 390 ppmv [42]. Usually bimolecular ion-molecule reactions occurring at the collision rate ($\sim 3 \times 10^{-9}$ cm$^3\cdot$s^{-1}) limit analyte concentrations measured in a SIFT-MS instrument to less than 20 ppmv. Concentrations of analytes higher than this would require sample dilution. In the present case, however, an association rate coefficient of 3.1×10^{-28} cm$^6\cdot$molec$^{-2}\cdot$s^{-1} is equivalent to a binary rate coefficient of 3.2×10^{-12} cm$^3\cdot$s^{-1}. This rate coefficient is effectively three orders of magnitude less than a typical exoergic binary rate coefficient, allowing much higher concentrations of CO_2 to be measured. A rate coefficient of this magnitude allows linear changes with CO_2 concentrations up to 1000 ppmv for O^- (reaction 1) and over 40,000 ppmv for O_2^- (reaction 2) to be monitored. The rate coefficients for CO_2 have been previously determined [43].

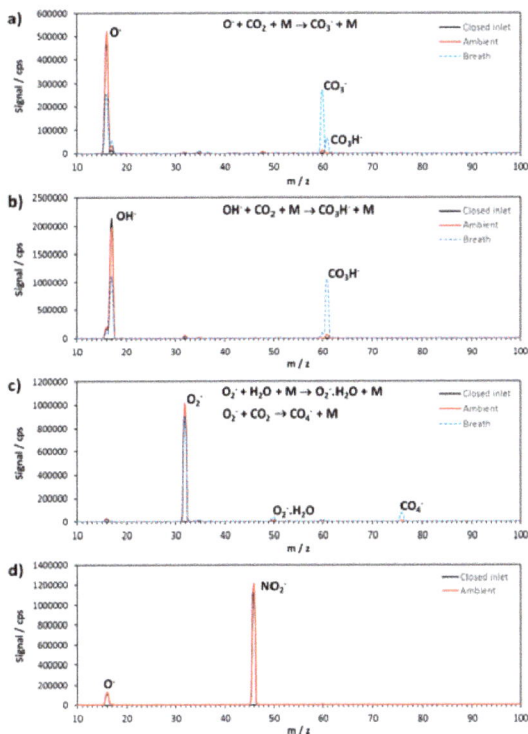

$$O^- + CO_2 + N_2 \rightarrow CO_3^- + N_2, \, k = 3.1 \times 10^{-28} \text{ cm}^6\cdot\text{molec}^{-2}\cdot\text{s}^{-1}, m/z = 60 \tag{1}$$

$$OH^- + CO_2 + N_2 \rightarrow HCO_3^- + N_2, \, k = 7.6 \times 10^{-28} \text{ cm}^6\cdot\text{molec}^{-2}\cdot\text{s}^{-1}, m/z = 61 \tag{2}$$

$$O_2^- + CO_2 + N_2 \rightarrow CO_4^- + N_2, \, k = 4.7 \times 10^{-29} \text{ cm}^6\cdot\text{molec}^{-2}\cdot\text{s}^{-1}, m/z = 76 \tag{3}$$

Figure 2. The downstream mass spectrometer traces for the reagent ions (**a**) O^-; (**b**) OH^-; (**c**) O_2^-; and (**d**) NO_2^- are shown under three configurations: with a closed sample inlet, only nitrogen is in the flow tube; the trace observed when an air sample is added is marked Ambient; and the trace of a human breath exhalation is marked Breath.

The linearity with concentrations of the two certified mixtures of CO_2 was checked using each reaction, and the results are shown in Figure 3a,b. Two additional secondary binary reactions that have some small contribution to the product ions in reactions (1) and (3) and that need to be included in the analysis as they also contribute to the magnitude of CO_3^- and CO_4^- are reactions (4) and (5) [44,45]:

$$O_3^- + CO_2 \rightarrow CO_3^- + O_2, k = 5.5 \times 10^{-10} \text{ cm}^3 \cdot \text{s}^{-1} \tag{4}$$

$$O_2^- \cdot H_2O + CO_2 \rightarrow CO_4^- + H_2O, k = 5.2 \times 10^{-10} \text{ cm}^3 \cdot \text{s}^{-1} \tag{5}$$

An interesting side effect of the concentration range for O_2^- and CO_2 is that it can be used to directly monitor CO_2 in breath, which typically ranges between 3% and 6% of a breath exhalation.

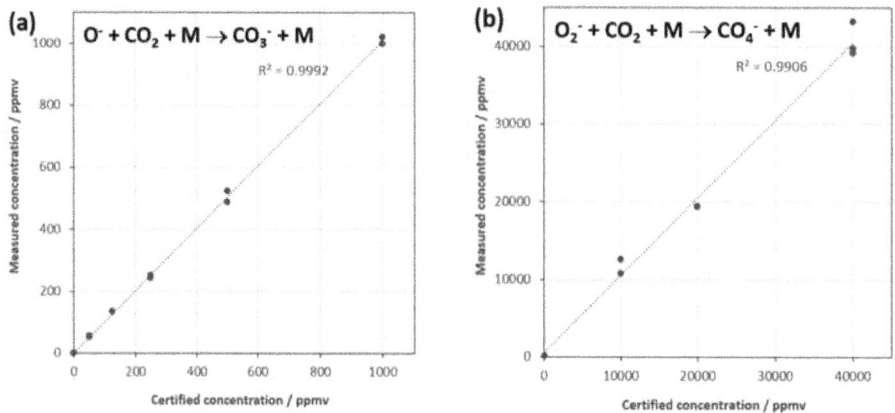

Figure 3. Measured concentrations of CO_2 when using the association reactions $O^- + CO_2 + M$ (a) and the $O_2^- + CO_2 + M$ reaction (b).

3.1. Reactions with Fumigants

Vikane: Because of their rapid response time and ease of use, SIFT-MS instruments have been used widely by Border Protection Agencies and Contract Testing Companies to monitor shipping containers that have been fumigated. One of these fumigant chemicals is Vikane (sulfuryl fluoride) or SO_2F_2. SO_2F_2 is not reactive with the three positive reagent ions, but does exhibit rapid reactions with O^-, OH^-, and O_2^-.

$$O^- + SO_2F_2 \rightarrow SO_3F^- + F, k \approx 3.0 \times 10^{-10} \text{ cm}^3 \cdot \text{s}^{-1} \tag{6}$$

$$OH^- + SO_2F_2 \rightarrow SO_3F^- + HF, k \approx 5.4 \times 10^{-10} \text{ cm}^3 \cdot \text{s}^{-1} \tag{7}$$

$$O_2^- + SO_2F_2 \rightarrow SO_2F_2^- + O_2, k \approx 5.7 \times 10^{-10} \text{ cm}^3 \cdot \text{s}^{-1} \tag{8}$$

These reagent ions give excellent concentration data, as is demonstrated in Figure 4 for reaction of the OH^- reagent ion with Vikane in comparison with static dilutions of the certified mixture of Vikane (5.02 ppmv $\pm$ 2%) in air.

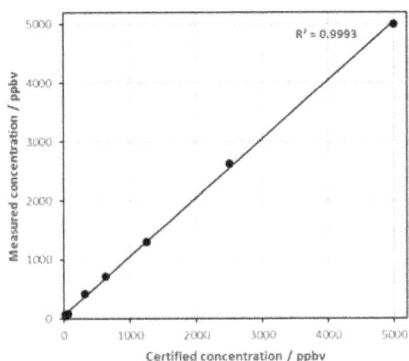

Figure 4. Comparison of SIFT-MS measurement against various concentrations of Vikane obtained by static dilutions of a certified 5.02 ppmv Vikane/air mixture.

3.2. Other Fumigants

A summary of the reaction chemistry observed for other fumigants examined with the negative reagent ions are presented in Table 1 (O^-), Table 2 (OH^-), and Table 3 ($O_2{}^-$). In these tables, the rate coefficients that have been estimated in the present work for chloropicrin and phosphine have been based on the certified concentrations of these fumigants in permeation tubes in the low ppmv range. These concentrations were also confirmed by the known reactions of these fumigants with the positively charged reagent ions H_3O^+ or $O_2{}^+$. Based on these known data, the rate coefficients for chloropicrin and phosphine were adjusted for the negative ion reagents to give the same values as for the positive reagent ions, and are presented as approximate by the ~symbol. The branching ratio column refers to a reaction of the reagent ion with the analyte in which multiple product ions with the fumigant are formed. It represents the fraction of reactions that terminate in the stated product ion.

Table 1. O^- reactions of fumigants, nitrogen carrier gas, and flow tube at 120 °C.

Fumigant	Ion Product	Branching Ratio	Rate Coefficient/cm^3·s^{-1}
Chloropicrin	$CCl_2NO_2{}^-$	1.0	~1.6×10^{-9}
Hydrogen cyanide	CN^- CNO^-	~0.9 ~0.1	3.7×10^{-9} [a]
Methyl chloride	OH^- Cl^- $CClH^-$	0.45 0.40 0.15	1.7×10^{-9} [b,c]
Phosphine	$PH_2{}^-$ PH_2O^-	0.6 0.4	~5.0×10^{-10}
Dibromoethane	Br^-	1.0	2.2×10^{-9} [d]

[a] Reference [46]; [b] Reference [47]; [c] Reference [48]; [d] Reference [49].

Table 2. OH^- reactions of fumigants in nitrogen carrier gas and flow tube at 120 °C.

Fumigant	Ion Product	Branching Ratio	Rate Coefficient/cm^3·s^{-1}
Chloropicrin	$CCl_2NO_2{}^-$	1.0	~1.5×10^{-9}
Hydrogen cyanide	CN^-	1.0	3.5×10^{-9} [a]
Methyl chloride	Cl^-	1.0	1.5×10^{-9} [b]
Phosphine	$PH_2{}^-$	1.0	~1.0×10^{-9}
Dibromoethane	Br^-	1.0	2.2×10^{-9} [c]

[a] Reference [50]; [b] Reference [47]; [c] Reference [49].

Table 3. O_2^- reactions of fumigants in nitrogen carrier gas and flow tube at 120 °C.

Fumigant	Ion Product	Branching Ratio	Rate Coefficient/cm$^3 \cdot$s^{-1}
Chloropicrin	no reaction		
Hydrogen cyanide	CN$^-$	1.0	~3.5 × 10^{-9}
Methyl chloride	Cl$^-$	1.0	7.4 × 10^{-10} [a,b]
Phosphine	no reaction		
Dibromoethane	Br$^-$	1.0	1.9 × 10^{-9} [b]

[a] Reference [51]; [b] Reference [48].

3.3. Acid Gases

Most of the acid gases are unreactive with the three positively charged reagent ions commonly used in SIFT-MS instruments. The present expansion of the reagent ions used in SIFT-MS to the negative ions O^-, OH^-, O_2^-, NO_2^-, and NO_3^- has made monitoring of these acid gases relatively simple using the direct analysis technique of SIFT-MS. The ion chemistry is summarised in Tables 4–6, and known concentrations were investigated using calibrated mixtures of each analyte in nitrogen (unless specified otherwise), as discussed previously.

Table 4. O^- reactions of acid gases, nitrogen carrier gas, and flow tube at 120 °C.

Acid Gas	Ion Product	Branching Ratio	Rate Coefficient/cm$^3 \cdot$s^{-1}
Hydrogen fluoride	F$^-$	1.0	5.0 × 10^{-10} [a]
Hydrogen chloride	Cl$^-$	~0.95	1.3 × 10^{-9} [b]
Sufur dioxide	mainly electron detachment		2.1 × 10^{-9} [c]
Sulfur trioxide	SO$_3^-$	0.87	1.8 × 10^{-9} [d]
	SO$_2^-$	0.13	
Hydrogen sulfide	SH$^-$	1.0	~2.5 × 10^{-10}
Nitrogen dioxide	NO$_2^-$	1.0	1.0 × 10^{-9} [e]

[a] Reference [52]; [b] Reference [53]; [c] Reference [54]; [d] Reference [55]; [e] Reference [56].

Table 5. OH^- reactions of acid gases, nitrogen carrier gas, and flow tube at 120 °C.

Acid Gas	Ion Product	Branching Ratio	Rate Coefficient/cm$^3 \cdot$s^{-1}
Hydrogen fluoride	F$^-$	~1.0	2.75 × 10^{-9} [a]
Hydrogen chloride	Cl$^-$	1.0	~1.3 × 10^{-9}
Sufur dioxide	OH$^-$ SO$_2$	1.0	1.0 × 10^{-26} [b]
Sulfur trioxide	SO$_3^-$	1.0	1.6 × 10^{-9} [c]
Hydrogen sulfide	SH$^-$	1.0	~7 × 10^{-10}
Nitrogen dioxide	NO$_2^-$	1.0	1.1 × 10^{-9} [d]

[a] Reference [57]; [b] Reference [43] (units are cm$^6 \cdot$molecule$^{-2} \cdot$s^{-1}); [c] Reference [55]; [d] Reference [58].

Table 6. O_2^- reactions of acid gases, nitrogen carrier gas, and flow tube at 120 °C.

Acid Gas	Ion Product	Branching Ratio	Rate Coefficient/cm$^3 \cdot$s^{-1}
Hydrogen fluoride	F$^-$	~1.0	Not measured
Hydrogen chloride	Cl$^-$	~1.0	1.2 × 10^{-9} [a]
Sufur dioxide	SO$_2^-$	1.0	1.9 × 10^{-9} [b]
Sulfur trioxide	SO$_3^-$	1.0	1.5 × 10^{-9} [c]
Hydrogen sulfide	SH$^-$	1.0	~1.3 × 10^{-9}
Nitrogen dioxide	NO$_2^-$	1.0	7.0 × 10^{-10} [d]

[a] Reference [53]; [b] Reference [45]; [c] Reference [55]; [d] Reference [56].

Perhaps the biggest problem in monitoring the most reactive acid gases such as HF using a direct technique was having an inlet system conditioned to the transmission of HF without loss of HF to the walls of the inlet tubing. In this test, heated $\frac{1}{4}$ inch OD polytetrafluoroethylene tubing was used to transport a calibrated mixture (10 ppmv) into the instrument. Inlet conditioning times of at least three hours were required before the HF concentration stabilized and monitored concentrations approached the manufacturers' levels. Once the conditioning process was completed, the HF measurements can be completed in a few seconds.

3.4. Other Negatively Charged Reagent Ions

In the brief discussion outlined here on the application of negative reagent ions with various analytes, we have not discussed much of the ion-molecule chemistry of NO_2^- and NO_3^-. These two ions are also produced from a microwave discharge of dry air in the instrument ion source with air as the source gas. The NO_3^- ion is largely unreactive with many of the common environmental contaminants but the NO_2^- ion, although less reactive than the negative ions already discussed, does undergo some useful reactions with some analytes. One example is the reaction of NO_2^- and HCl, which provides a very useful measure of the HCl concentration.

The reaction between NO_2^- and HCl is [59]

$$NO_2^- + HCl \rightarrow Cl^- + HNO_2, k = 1.4 \times 10^{-9}\ cm^3 \cdot s^{-1} \tag{9}$$

A linear relationship is found using a SIFT-MS instrument by monitoring the ratio of the Cl^- product ion and the NO_2^- reagent ion for concentrations of HCl from ppbv ranging up to 35 ppmv, and is shown in Figure 5.

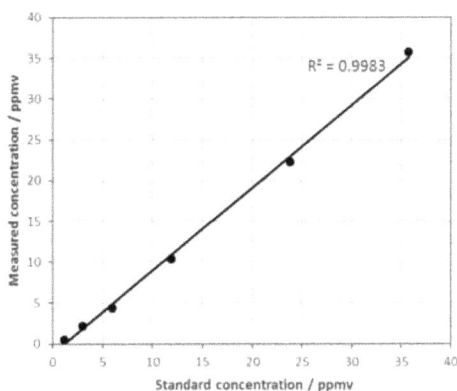

Figure 5. Comparison of a SIFT-MS measurement against the concentration of a certified mixture of HCl in nitrogen using the NO_2^- reagent ion.

Finally, we note that in all the analytes mentioned in this work in Tables 1–6, linear correlations were found with concentration using nitrogen as the carrier flow gas, further underscoring the usefulness of the SIFT-MS technique by extending the carrier gas from helium to nitrogen. We also note that when the change from a helium to a nitrogen carrier gas is implemented, enhanced association reactions may occur such as the reactions coverting H_3O^+ to $H_3O^+ \cdot (H_2O)_n$ where $n = 1$–3.

4. Conclusions

In the past few years, direct analytical techniques utilizing mass spectrometric detection have greatly simplified the process of monitoring analytes at trace levels in air samples. SIFT-MS, in particular, because of the selection of reagent ions readily available to the technique and the ease of transitioning between them, has proven very effective at monitoring a wide range of analytes in the areas of medicine, the environment, and food and flavor chemistry. The analyte concentration is simply found from the ratio of the product ion counts to the reagent ion counts, the known flow conditions, and an instrument calibration factor [4].

One constraint to the number of applications previously available to SIFT-MS was that not all analytes were accessible to the technique. Several smaller analyte molecules in areas such as fumigation and those of the acid gases were unreactive with the positively charged reagent ions used. In this work, we have extended the number of reagent ions to negatively charged ions, and in the process generated five new negatively charged ions (O^-, OH^-, O_2^-, NO_2^-, and NO_3^-) that may be used to react with analyte molecules. We have demonstrated their effectiveness at monitoring selected analytes that were formerly blind to positive reagent ions over a wide concentration range. We have also demonstrated the effectiveness of using nitrogen as a carrier gas for the SIFT-MS flow tube reactor in comparison with helium.

Acknowledgments: We thank Thomas Hughes and Nic Lamont for making preliminary measurements on some of these reactions.

Author Contributions: Vaughan S. Langford and Daniel B. Milligan conceived the range of experiments designated to show the difference between positive and negative ion-analyte chemistry; David Hera and Thomas I. McKellar performed and monitored the experiments reported; Murray J. McEwan assisted with the process of generating negative ions and wrote the paper.

Conflicts of Interest: The authors declare there were no conflicts of interest.

References

1. Watson, J.J.; Sparkman, O.D. *Introduction to Mass Spectrometry*, 4th Ed. ed; John Wiley & Sons, Ltd.: Hoboken, NJ, USA, 2008.
2. McEwan, M.J. Direct analysis mass spectrometry. In *Ion/Molecule Attachment Reactions*; Fujii, T., Ed.; Springer Science & Business Media: New York, NY, USA, 2015.
3. Spanel, P.; Smith, D. Selected ion flow tube: A technique for quantitative trace gas analysis of air and breath. *Med. Biol. Eng. Comput.* **1996**, *34*, 409–419. [CrossRef] [PubMed]
4. Smith, D.; Spanel, P. Selected ion flow tube mass spectrometry (SIFT-MS) for on-line trace gas analysis. *Mass Spec. Rev.* **2005**, *24*, 661–700. [CrossRef] [PubMed]
5. Spanel, P.; Smith, D. Progress in SIFT-MS: Breath analysis and other applications. *Mass Spectrom. Rev.* **2011**, *30*, 236–267. [CrossRef] [PubMed]
6. Smith, D.; Spanel, P.; Herbig, J.; Beauchamp, J. Mass Spectrometry for real time quantitative breath analysis. *J. Breath Res.* **2014**, *8*, 027101. [CrossRef] [PubMed]
7. Storer, M.; Curry, K.; Squire, M.; Kingham, S.; Epton, M. Breath testing and personal exposure—SIFT-MS detection of breath acetonitrile for exposure monitoring. *J. Breath Res.* **2015**, *9*, 036006. [CrossRef] [PubMed]
8. Turner, C.; Spanel, P.; Smith, D. A longitudinal study of breath isoprene in healthy volunteers using selected ion flow tube mass spectrometry (SIFT-MS). *Physiol. Meas.* **2006**, *27*, 13–22. [CrossRef] [PubMed]
9. Pysanenko, A.; Spanel, P.; Smith, D. A study of sulfur-containing compounds in mouth-and nose-exhaled breath and in the oral cavity using selected ion flow tube mass spectrometry. *J. Breath Res.* **2008**, *2*, 046004. [CrossRef] [PubMed]
10. Kumar, S.; Huang, J.; Hanna, G.B. SIFT-MS analysis of headspace vapour from gastric content for the diagnosis of gastro-oesophageal cancer. *Br. J. Surg.* **2013**, *100*, S4.
11. Zeft, A.; Costanzo, D.; Alkhouri, N.; Patel, N.; Grove, D.; Spalding, S.J.; Dweik, R. Metabolomic analysis of breath volatile organic compounds reveals unique breathprints in children with juvenile idiopathic arthritis. *Arthritis Rheumatol.* **2014**, *66*, S159. [CrossRef]
12. Alkhouri, N.; Eng, K.; Cikach, F.; Patel, N.; Yan, C.; Brindle, A.; Rome, E.; Hanouneh, I.; Grove, D.; Lopez, R.; et al. Breathprints of childhood obesity: Changes in volatile organic compounds in obese children compared with lean controls. *Pediatri. Obes.* **2015**, *10*, 23–29. [CrossRef] [PubMed]

13. Patel, N.; Alkhouri, N.; Eng, K.; Cikach, F.; Mahajan, L.; Yan, C.; Grove, D.; Rome, E.S.; Lopez, R.; Dweik, R.A. Metabolomic analysis of breath volatile organic compounds reveal unique breathprints in children with inflammatory bowel disease: A pilot study. *Aliment. Pharmacol. Ther.* **2014**, *40*, 498–506. [PubMed]

14. Boshier, P.R.; Mistry, V.; Cushnir, J.R.; Kon, O.M.; Elkin, S.L.; Curtis, S.; Marczin, N.; Hanna, G.B. Breath metabolite response to major upper gastrointestinal surgery. *J. Surg. Res.* **2015**, *193*, 704–712. [CrossRef] [PubMed]

15. Navaneethan, U.; Parsi, M.A.; Lourdusamy, V.; Bhatt, A.; Gutierrez, N.G.; Grove, D.; Sanaka, M.R.; Hammel, J.P.; Stevens, T.; Vargo, J.J. Volatile organic compounds in bile for early diagnosis of cholangiocarcinoma in patients with primary sclerosing cholangitis: A pilot study. *Gastrointest. Endosc.* **2015**, *81*, 943. [CrossRef] [PubMed]

16. Michalcikova, R.; Dryahina, K.; Spanel, P. SIFT-MS quantification of several breath biomarkers of inflammatory bowel disease, IBD: A detailed study of the ion chemistry. *Int. J. Mass Spectrom.* **2016**, *396*, 35–41. [CrossRef]

17. Langford, V.S.; Gray, J.D.C.; Maclagan, R.G.A.R.; McEwan, M.J. Detection of siloxanes in landfill gas and biogas using SIFT-MS. *Curr. Anal. Chem.* **2013**, *9*, 558–565. [CrossRef]

18. Prince, B.J.; Milligan, D.B.; McEwan, M.J. Application of selected ion flow tube mass spectrometry to real-time atmospheric monitoring. *Rapid Commun. Mass Spectrom.* **2010**, *24*, 1763–1769. [CrossRef] [PubMed]

19. Hastie, D.R.; Gray, J.; Langford, V.S.; Maclagan, R.G.A.R.; Milligan, D.B.; McEwan, M.J. Real-time measurement of peroxyacetal nitrate using SIFT-MS. *Rapid Commun. Mass Spectrom.* **2010**, *24*, 343–348. [CrossRef] [PubMed]

20. Volckaert, D.; Ebude, D.E.L.; Van Langenhove, H. SIFT-MS analysis of the removal of dimethyl sulphide, n-hexanone and toluene from waste air by a two phase partitioning bioreactor. *Chem. Eng. J.* **2016**, *290*, 346–352. [CrossRef]

21. Van Huffel, K.; Heynderickx, P.M.; Dewulf, J.; Van Langenhove, H. Measurement of odorants in livestock buildings: SIFT-MS and GC-MS. *Chem. Eng. Trans.* **2012**, *30*, 67–72.

22. Sovova, K.; Shestivska, V.; Spanel, P. Real-time quantification of traces of biogenic volatiles selenium compounds is humid air by SIFT-MS. *Anal. Chem.* **2012**, *84*, 4979–4983. [CrossRef] [PubMed]

23. Volckaert, D.; Heynderickx, P.M.; Fathi, E.; Van Langenhove, H. SIFT-MS: A novel tool for monitoring and evaluating a biofilter performance. *Chem. Eng. J.* **2016**, *304*, 98–105. [CrossRef]

24. Romanias, M.N.; Ourrad, H.; Thevenet, F.; Riffault, V. Investigating the heterogeneous interaction of VOCs with natural atmospheric particles: Adsorption of limonene and toluene on Saharan mineral dust. *J. Phys. Chem. A* **2016**, *120*, 1197–1212. [CrossRef] [PubMed]

25. Storer, M.; Salmond, J.; Dirks, K.N.; Kingham, S.; Epton, M. Mobile selected ion flow tube mass spectrometry (SIFT-MS) devices and their use for pollution exposure monitoring in breath and ambient air-pilot study. *J. Breath Res.* **2014**, *8*, 037106. [CrossRef] [PubMed]

26. Francis, G.J.; Langford, V.S.; Milligan, D.B.; McEwan, M.J. Real time monitoring of hazardous air pollutants. *Anal. Chem.* **2009**, *81*, 1595–1599. [CrossRef] [PubMed]

27. Xu, Y.; Barringer, S. Effect of temperature on lipid-related volatile production in tomato puree. *J. Agric. Food Chem.* **2009**, *57*, 9108–9133. [CrossRef] [PubMed]

28. Hansanugram, A.; Barringer, S.A. Effect of milk on the deodorization of malodorous breath after garlic ingestion. *J. Food Sci.* **2010**, *75*, C549–C558. [CrossRef] [PubMed]

29. Huang, Y.; Barringer, S.A. Alkyl pyrazines and other volatiles in cocoa liquors at pH 5 to 8 by SIFT-MS. *J. Food Sci.* **2010**, *75*, C121–C127. [CrossRef] [PubMed]

30. Davis, B.M.; Senthilmohan, S.T.; McEwan, M.J. Direct determination of antioxidants in whole olive oil using the SIFT-MS-TOSC assay. *J. Am. Oil Chem. Soc.* **2011**, *88*, 785–792. [CrossRef]

31. Huang, Y.; Barringer, S.A. Monitoring cocoa volatiles produced during roasting by SIFT-MS. *J. Food Sci.* **2011**, *76*, C279–C286. [CrossRef] [PubMed]

32. Langford, V.S.; Reed, C.J.; Milligan, D.B.; McEwan, M.J.; Barringer, S.A.; Harper, J. Headspace analysis of Italian and New Zealand Parmesan Cheese. *J. Food Sci.* **2012**, *77*, C719–C726. [CrossRef] [PubMed]

33. Noseda, B.; Islam, M.T.; Eriksson, M.; Heyndrickx, M.; De Reu, K.; Van Langenhove, H.; Devlieghere, F. Microbiological spoilage of vacuum and modified packaged Vietnamese Pangasius hypophthalmus fillets. *Food Microbiol.* **2012**, *30*, 408–419. [CrossRef] [PubMed]

34. Langford, V.S.; Gray, J.; Foulkes, B.; Bray, P.; McEwan, M.J. Application of SIFT-MS to the characterization of monofloral New Zealand honeys. *J. Agric. Food Chem.* **2012**, *60*, 6806–6815. [CrossRef] [PubMed]

35. Munch, R.; Barringer, S.A. Deodorization of garlic breath volatiles by food and food components. *J. Food Sci.* **2014**, *79*, C526–C533. [CrossRef] [PubMed]

36. Huang, Y.; Barringer, S.A. Kinetics of furan formation during pasteurization of soy source. *LWT Food Sci. Technol.* **2016**, *67*, 200–205.

37. Castada, H.Z.; Wick, C.; Harper, J.W.; Barringer, S. Headspace quantification of pure and aqueous solutions of binary mixtures of key volatile organic compounds in Swiss cheese using selected ion flow tube mass spectrometry. *Rapid Commun. Mass Spectrom.* **2015**, *29*, 81–90. [CrossRef] [PubMed]

38. Langford, V.S.; Graves, I.; McEwan, M.J. Rapid monitoring of volatile organic compounds: A comparison between GC/MS and SIFT-MS. *Rapid Commun. Mass Spectrom.* **2014**, *28*, 10–18. [CrossRef] [PubMed]

39. Milligan, D.B.; Fairley, D.A.; Freeman, C.G.; McEwan, M.J. A flowing afterglow selected ion flow tube (FA/SIFT) comparison of SIFT injector flanges and $H_3^+ + N$ revisited. *Int. J. Mass Spectrom.* **2000**, *202*, 351–361. [CrossRef]

40. Milligan, D.B.; Francis, G.J.; Prince, B.J.; McEwan, M.J. Demonstration of SIFT-MS in the parts per trillion range. *Anal. Chem.* **2007**, *79*, 2537–2540. [CrossRef] [PubMed]

41. Spanel, P.; Dryahina, K.; Smith, D. Microwave plasma ion sources for selected ion flow tube mass spectrometry: Optimizing their performance and detection limits for trace detection. *Int. J. Mass Spectrom.* **2007**, *267*, 117–124. [CrossRef]

42. Pachauri, R.K.; Meywer, L.A. (Eds.) Contribution of working groups I, II and III to the fifth assessment report of the intergovernmental panel on climate change. In *IPCC 2014 Climate Change: Synthesis Report*; IPCC: Geneva, Switzerland, 2014.

43. Fehsenfeld, F.C.; Ferguson, E.E. Laboratory studies of negative ion reactions with atmospheric trace constituents. *J. Chem. Phys.* **1974**, *61*, 3181–3193. [CrossRef]

44. Dotan, I.; Davidson, J.A.; Streit, G.E.; Albritton, D.L.; Fehsenfeld, F.C. A study of the reaction of $O_3^- + CO_2$ and its implications on the thermochemistry of CO_3 and O_3 and their negative ions. *J. Chem. Phys.* **1977**, *67*, 2874–2879. [CrossRef]

45. Fahey, D.W.; Bohringer, H.; Fehsenfeld, F.C.; Ferguson, E.E. Reaction rate constants for $O_2^-(H_2O)_n$ ions $n = 0$–4 with O_3, NO, SO_2 and CO_2. *J. Chem. Phys.* **1982**, *76*, 1799–1805. [CrossRef]

46. Bohme, D.K. The kinetics and energetics of proton transfer reactions. In *Interactions between Ions and Molecules*; Plenum Press: New York, NY, USA, 1975; p. 489.

47. Tanaka, K.; Mackay, G.I.; Payzant, J.D.; Bohme, D.K. Gas phase reactions of anions with halogenated methanes 297 ± 2K. *Can. J. Chem.* **1976**, *54*, 1643–1659. [CrossRef]

48. Mayhew, C.A.; Peverall, R.; Watts, P. Gas phase ionic reactions of freons and related compounds: Reactions of some halogenated methanes with O^- and O_2^-. *Int. J. Mass Spectrom. Ion Proc.* **1993**, *125*, 81–93. [CrossRef]

49. Thomas, R.; Liu, Y.; Mayhew, C.A.; Peverall, R. Selected ion flow tube studies of the gas pahse reactions of O^-, O_2^- and OH^- with a variety of brominated compounds. *Int. J. Mass Spectrom. Ion Proc.* **1996**, *155*, 163–183. [CrossRef]

50. Raksit, A.B.; Bohme, D.K. An experimental study of the influence of hydration on the reactivity of the hydroxide anion in the gas phase at room temperature. *Can. J. Chem.* **1983**, *61*, 1683–1689. [CrossRef]

51. McDonald, R.N.; Chowdhury, A.K. Gas phase ion-molecule reactions of dioxygen anion radical (O_2^-). *J. Am. Chem. Soc.* **1985**, *107*, 4123–4128. [CrossRef]

52. Hamilton, C.E.; Duncan, M.A.; Zwier, T.S.; Weisshaar, J.V.; Ellison, G.B.; Bierbaum, V.M.; Leone, S.R. Product vibrational analysis of ion molecule reactions by laser induced fluorescence in a flowing afterglow. $O^- + HF \rightarrow OH^- (v = 0.1) + F^-$. *Chem. Phys. Lett.* **1983**, *94*, 4–9. [CrossRef]

53. Streit, G.E. Gas phase reactions of O^- and O_2^- with a variety of halogenated compounds. *J. Phys. Chem.* **1982**, *86*, 2321–2324. [CrossRef]

54. Lindinger, W.; Albritton, D.L.; Fehsenfeld, F.C.; Ferguson, E.E. Reactions of O^- with N_2, N_2O, SO_2, NH_3, CH_4 and C_2H_4 and $C_2H_2^-$ with O_2 from 300K to relative kinetic energies of ~2 eV. *J. Chem. Phys.* **1975**, *63*, 3238–3242. [CrossRef]

55. Arnold, S.T.; Morris, R.A.; Viggiano, A.A.; Jayne, J.T. Ion chemistry relevant for chemical detection of SO_3. *J. Geophys. Res.* **1995**, *100*, 14141–14146. [CrossRef]

56. Ikezoe, Y.; Viggiano, A. *Gas Phase Ion-Molecule Reaction Rate Constants through 1986*; Ion Reaction Research Group of the Mass Spectrometry Society of Japan: Tokyo, Japan, 1987.

57. Hierl, P.M.; Ahrens, A.F.; Henchman, M.; Viggiano, A.A.; Paulson, J.F. Proton transfer as a function of hydration number and temperature: Rate constants and product distributions for $OH^-(H_2O)_{0-3}$ + HF at 200K–500K. *J. Am. Chem. Soc.* **1986**, *108*, 3140–3142. [CrossRef]

58. Tanner, S.D.; Mackay, G.I.; Bohme, D.K. An experimental study of the reactivity of the hydroxide anion in the gas phase at room temperature and its perturbation by hydration. *Can. J. Chem.* **1981**, *59*, 1615–1621. [CrossRef]

59. Ferguson, E.E.; Dunkin, D.B.; Fehsenfeld, F.C. Reactions of NO_2^- and NO_3^- with HCl and HBr. *J. Chem. Phys.* **1972**, *57*, 1459–1463. [CrossRef]

environments

MDPI

Article

Highly Sensitive and Selective VOC Sensor Systems Based on Semiconductor Gas Sensors: How to?

Andreas Schütze [1,*], Tobias Baur [1], Martin Leidinger [1], Wolfhard Reimringer [2], Ralf Jung [1], Thorsten Conrad [2] and Tilman Sauerwald [1]

[1]   Lab for Measurement Technology, Department of Systems Engineering, Saarland University, Saarbrücken 66041, Germany; t.baur@lmt.uni-saarland.de (T.B.); m.leidinger@lmt.uni-saarland.de (M.L.); ralfjung@gmx.net (R.J.); t.sauerwald@lmt.uni-saarland.de (T.S.)

[2]   3S GmbH—Sensors, Signal Processing, Systems, Saarbrücken 66121, Germany; reimringer@3s-ing.de (W.R.); conrad@3s-ing.de (T.C.)

*   Correspondence: schuetze@lmt.uni-saarland.de; Tel.: +49-681-302-4664

Academic Editors: Ki-Hyun Kim and Abderrahim Lakhouit
Received: 31 December 2016; Accepted: 24 February 2017; Published: 1 March 2017

Abstract: Monitoring of volatile organic compounds (VOCs) is of increasing importance in many application fields such as environmental monitoring, indoor air quality, industrial safety, fire detection, and health applications. The challenges in all of these applications are the wide variety and low concentrations of target molecules combined with the complex matrix containing many inorganic and organic interferents. This paper will give an overview over the application fields and address the requirements, pitfalls, and possible solutions for using low-cost sensor systems for VOC monitoring. The focus lies on highly sensitive metal oxide semiconductor gas sensors, which show very high sensitivity, but normally lack selectivity required for targeting relevant VOC monitoring applications. In addition to providing an overview of methods to increase the selectivity, especially virtual multisensors achieved with dynamic operation, and boost the sensitivity further via novel pro-concentrator concepts, we will also address the requirement for high-performance gas test systems, advanced solutions for operating and read-out electronic, and, finally, a cost-efficient factory and on-site calibration. The various methods will be primarily discussed in the context of requirements for monitoring of indoor air quality, but can equally be applied for environmental monitoring and other fields.

Keywords: low-cost sensors; VOC sensor systems; sensitivity; selectivity; virtual multisensor; calibration

1. Introduction

Measurements of volatile organic compounds (VOCs) are becoming ever more important due to stringent environmental regulations and increasing health concerns. Typical compounds with high relevance are benzene, naphthalene, formaldehyde, and tetrachloroethylene, but there are many more depending on the specific environment and the target application. Exposure to VOCs for a long time can have negative effects on human health, including damage to the respiratory system and skin irritations [1]. Moreover, VOCs are the main cause of the sick building syndrome [2,3]. Besides these unspecific adverse health effects, some VOCs are proven to be carcinogenic (e.g., benzene [4]) or are suspected to be carcinogenic (e.g., formaldehyde [5]). The specific challenge for VOC measurements are the low target concentrations: the respective guideline threshold values for some critical substances (in indoor air) are 0.1 mg/m^3 (81 ppb) for formaldehyde and 0.01 mg/m^3 (1.9 ppb) for naphthalene according to the World Health Organization (WHO) [4], and 5 µg/m^3 (1.6 ppb) for benzene according to EU guidelines [6]. In other fields such as industrial monitoring and workplace safety, higher values apply, but these are currently trending down sharply, e.g., in Germany [7]. Furthermore, very high selectivity is required to discriminate highly toxic or carcinogenic VOCs from inorganic gases such as carbon monoxide (CO), nitrous oxides (NO_x), ozone (O_3), or hydrogen (H_2), all of which occur naturally or are generated by (other) pollution sources less toxic or from benign VOCs such as ethanol. For the total VOC concentration (TVOC), WHO suggests a limit value of 1 ppm, i.e., three orders of magnitude above the lowest threshold limit values for hazardous VOCs. In other applications, the ratio between target VOC and interferent can even reach five orders of magnitude, e.g., for smoldering fire detection on coal mines, where 100 ppb ethane (C_2H_4) should be detected against a background of up to 1% methane (CH_4) [8]. A field that has gained increasing interest in recent years in monitoring of odor compounds, which, while often not having a direct health effect, nevertheless considerably impact our quality of life. Typical compounds in this context are isovaleric acid ("sweat odor"), organosulfur compounds (thiols or mercaptan, e.g., (di-)methyl sulfide) as well as many esters, terpenes, amines, ketones, and, of course, aromatic compounds as well as hydrogen sulfide as an important inorganic odorant. Odor monitoring plays an increasing role in indoor air quality, but also in outdoor environmental monitoring, both at the source (emission measurements at the stack or the fence line) and at the impact side (odor nuisance monitoring) for various industries, e.g., waste treatment. Figure 1 gives an overview over VOC monitoring applications, also indicating relevant target gases and interferents. A more comprehensive overview is given in [9].

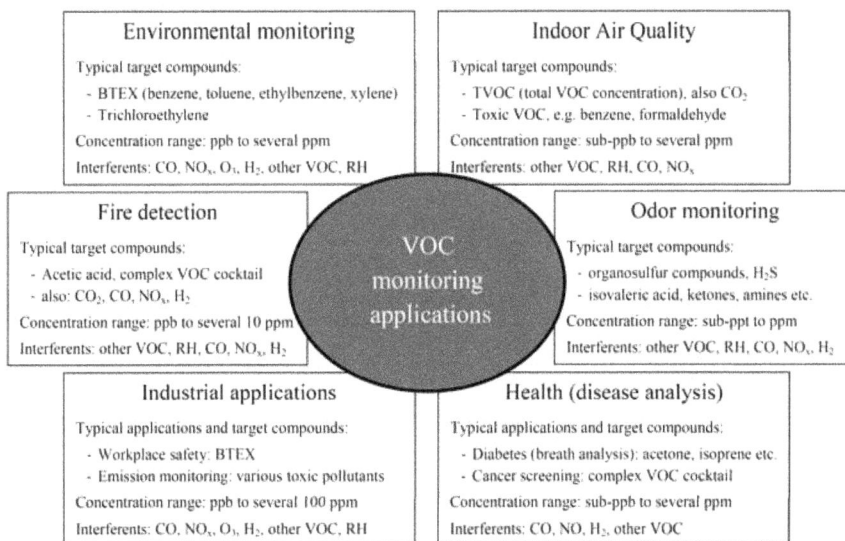

Figure 1. Overview over various applications requiring monitoring of VOCs plus typical target gases and relevant concentration ranges. Due to the wide variety of gases and applications in the specific fields, this is indicative only. Common challenges for most applications are the low target concentrations and the complex matrix. Note that in some cases the same compounds can be seen as target and interfering gases due to different sources.

Very few monitoring techniques actually achieve ppb (or even sub-ppb) level sensitivity so that sampling techniques are generally used for VOC monitoring. The standard method for VOC monitoring is sampling, e.g., with Tenax® as absorbent material, and then releasing the sampled gas into a gas chromatograph followed by identification of the VOCs by mass spectrometry (GC-MS). This method, while widely accepted, does however pose serious problems for comprehensive VOC monitoring as both very volatile organic compounds (VVOCs) and semivolatile organic compounds (SVOCs) will not be sampled with the standard technique [10]. VVOCs will simply pass through the sampling material, while SVCOCs are not released due to their very high boiling point. One prominent example for this shortcoming is formaldehyde for which a specific separate sampling and quantification protocol is required [11]. On the other hand, weak sorbents are required for trapping SVOCs. This means that standard measurement techniques are actually blind to a fairly wide range of compounds and that our understanding of environmental pollution and health effects is somewhat limited due to missing data on these compounds. Research is ongoing in finding suitable sorbent materials, e.g., multibed focusing traps, to allow the acquisition of more complete information on VOCs [12]. In addition, sampling only allows determining time-weighted average (TWA) values either for the long term (1 h to 24 h) or the short term (5 min to 60 min), thus possibly missing relevant short concentration peaks. Sensor-based monitoring, on the other hand, would allow both an improved temporal resolution and a wider detection spectrum, as sensors will respond to practically all VOCs including VVOCs and SVOCs.

2. Highly Sensitive Semiconductor Gas Sensor Principles

One candidate sensor principle for direct monitoring of even sub-ppb level VOC concentrations are metal oxide semiconductor (MOS) gas sensors [13,14], but gas-sensitive field effect transistors (GasFETs), e.g., based on silicon carbide (SiC-FET), have also proven their suitability for this task [15]. For simplification, we will concentrate on MOS sensors in the following discussion, but many aspects also apply to SiC-FET sensors, so the approach is presented for semiconductor gas sensors. Note that both sensor principles allow low-cost sensor solutions, at least for markets with high volumes. Even taking into account the advanced electronics and data processing that are necessary to achieve high sensitivity, selectivity, and stability, high performance sensor systems can be realized for well below 1.000€, as has been demonstrated for fire detection [16]. In mass applications, the system cost can drop to below 10€. Commercial screen-printed ceramic MOS sensors achieve detection limits down to sub-ppb levels due to the well-known grain boundary effect [17]. A small grain size down to nanocrystals [18] also improves the sensitivity as the bulk conductance of the material is not affected by gas adsorbed on the grain surface. Recently, nanostructured sensors based on nanowires [19] have been proposed and sometimes postulated as being necessary for achieving very low detection limits due to the high surface area achieved. Especially carbon nanotube-based sensors have been extensively studied with the expectation that these provide the key for very high sensitivity [20–22]. However, so far no rigorous study has proven that MOS nanowires or carbon nanotubes in fact achieve superior sensitivity compared to standard granular materials. Novel manufacturing methods for metal oxide layers, especially Pulsed Laser Deposition [23], achieve highly porous sensor layers with well-controlled morphology that can even reach high selectivity for some relevant target VOCs such as naphthalene [24].

On the other hand, we have recently developed a model that shows that the sensitivity of MOS sensors can be improved significantly by temperature cycling. The first benefit of varying the sensor temperature over a broad range is immediately obvious as the sensor will at some time in the cycle reach the optimal temperature, i.e., with the highest sensitivity, for the target gas(es). In addition, temperature cycling with fast temperature changes will lead to non-equilibrium surface conditions and thus reaches states that cannot be achieved in static operation at constant temperature. By carefully choosing the operating parameters, the sensitivity can actually be boosted considerably, e.g., by first operating the sensor at high temperature which leads to a high surface coverage with ionosorbed oxygen. If the sensor is then cooled down rapidly, this leads to a surface at low temperature but with an abundance of reactive oxygen ions—a condition that cannot be achieved in static mode. Without gas, the ionosorbed oxygen will slowly desorb to reach the equilibrium state with only few oxygen ions at low temperature. A gas interacting with the adsorbed oxygen will strongly influence this relaxation process leading to faster equilibration. The sensitivity and the respective sensor response G_{gas}/G_{zero} during dynamic operation have been shown to be orders of magnitude higher than in static operating mode [25]. Figure 2 describes this effect schematically. The non-equilibrium states during dynamic operation can also improve the selectivity, so optimized cycling can be used to address the three "S's"—sensitivity, selectivity, and stability—the key aspects for chemical sensor systems [26].

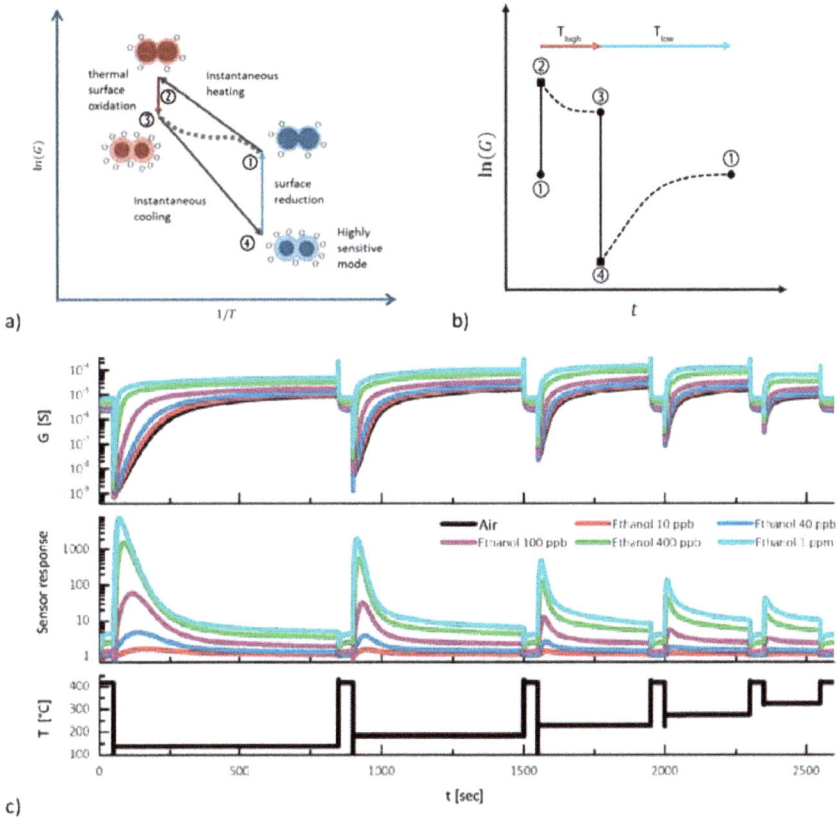

Figure 2. Boosting sensitivity of MOS sensors with temperature cycled operation (TCO): (a) MOS sensors show low oxygen coverage at low temperatures ① and high coverage at high temperatures ③ in steady state conditions (dashed line indicates equilibrium conditions between min and max temperature). Non-equilibrium surface states (②, ④) are achieved by fast temperature changes. Especially high oxygen coverage at low temperature will lead to a highly sensitive mode as target gases will react with the adsorbed oxygen. (b) Dynamic relaxation of the conductance after fast temperature changes boosts the sensitivity. (c) Admixture of VOCs leads to faster relaxation compared to pure air. Plotting the sensor response shows that a huge increase in sensitivity of several orders of magnitude can be achieved compared to the steady state response reached at the end of the relaxation process. The relaxation behavior at various temperatures is typical for the specific target gas and can thus be used to increase the selectivity of MOS-based sensor systems.

There are also other possibilities to enhance the response and increase the sensitivity of semiconductor gas sensors, especially optical excitation, which has been addressed by various groups [27,28]. Especially excitation with quantum energy above the bandgap energy of the semiconductor allows low-temperature operation and/or achieves higher sensitivity. In addition, light modulation can be used to extract signal patterns similar to TCO which can also increase the selectivity and stability of the overall sensor system.

3. Highly Selective Sensor Systems

The drawback of MOS (and SiC-FET) sensors is their inherent low selectivity, which is due to the sensor function principle. The high sensitivity of MOS sensors is due to the grain-boundary effect that is caused by ionosorbed oxygen leading to band bending at the surface of the grains and, thus, an energy barrier between grains with the conductance of the sensor being exponentially dependent on the height of this energy barrier. Any gas that either chemisorbs on the surface or interacts with the ionosorbed oxygen will change the energy barrier. Therefore, MOS sensors show a response to practically all relevant target and interfering gases except carbon dioxide (CO_2). To achieve the required high selectivity for gas discrimination and/or quantification of target gases in a background of interfering gases, several approaches are possible: (a) enhancing gas measurement systems with analytical tools, i.e., a GC tube to separate the various gas components (which will, however, suffer from similar drawbacks as sampling techniques for VVOC and SVOC); (b) making use of multisensor arrays and pattern recognition (often referred to as electronic nose) [9]; and (c) using dynamic operation, e.g., temperature cycling, to realize a virtual multisensor, which is also evaluated using typical pattern recognition methods [29]. The different methods employed for increasing the selectivity can also be combined, for example, by integrating several dynamically operated sensors in a hybrid sensor array or combining gas pre-concentration (see below) with temperature cycling to boost sensitivity and selectivity. Especially temperature cycled operation (TCO) has proven a very powerful and versatile tool for various sensor principles (MOS sensors [14,30–32], SiC-FETs [33], pellistors [34]), which is easily understandable as the chemical interaction between sensor and gas atmosphere is strongly influenced by the surface temperature. For example, some gases such as CO or H_2 will react at relatively low temperatures, while others such as CH_4 are more stable and thus require higher activation energies to cause a sensor reaction. Due to its simple implication and low cost (only some additional electronics are required), TCO is now widely accepted as a method to boost selectivity, especially as it also improves the stability of the sensor system due to self-cleaning of the sensor surface at higher temperatures [35] and the possibility to use features which are stable over time [32]. Furthermore, in addition to identifying and quantifying target gases, this approach allows sensor self-monitoring [36], which is a crucial aspect especially for applications in safety and security. For VOC monitoring, selective identification of hazardous VOCs down to concentrations of 1 ppb could be demonstrated even in a background of other VOCs of several ppm and against changing humidity using a commercial ceramic MOX sensor at least under lab conditions [14]; similarly, quantification of VOCs at low ppb levels was demonstrated with this approach both for MOS [26] and SiC-FET [15,37] sensors. Further approaches to improve the performance of semiconductor gas sensor systems with dynamic operation are optical excitation [27,28] and impedance spectroscopy [38] often applied to MOX sensors or gate bias cycled operation for GasFETs [39]. These approaches can also be combined to boost the selectivity further [40,41].

One often underestimated aspect for highly sensitive and selective sensor systems is the electronics for sensor operation and signal read-out. To make use of the full potential of dynamic operation, dedicated electronics that allow exact control of the sensor operating parameters and synchronized data acquisition are required. For temperature cycling, exact temperature control is key. Here, the heater integrated in each MOS (and SiC-FET) sensor is preferably also used as a temperature sensor to allow exact closed-loop control. This allows exact control even for highly dynamic cycles, especially for microstructured gas-sensors based on micro-hotplates, which exhibit thermal time constants of typically a few ms. This type of sensor is now manufactured at high volumes by various companies, e.g., ams Sensor Solutions, CCMOSS, Figaro, and SGX Sensortech, achieving very low-cost sensor elements. Appropriate electronics have been developed over several generations [42–44] and are today commercially available from 3S GmbH (3S Toolbox, [45]), which has been successfully employed also for outdoor odor nuisance monitoring [46]. To make full use of the TCO mode for MOS sensors, the sensor resistance has to be acquired with high temporal resolution (≥ 1 kHz) over a wide dynamic range of several orders of magnitude from kΩ (high gas concentrations at high operating temperatures) to GΩ or even TΩ (low gas concentration, non-equilibrium state after fast cool down, cf. Figure 2), ideally with constant relative resolution. The voltage across the sensor layer is limited, typically to ≤1 V, to prevent unwanted effects such as electromigration in the layer leading to sensor drift. However, this voltage limit (or more exactly the maximum field strength) also depends on the operating temperature, allowing higher voltages and thus more sensitive measurement at low temperature. Thus, electronics are required that allow fast and accurate measurement of very low currents in the pA range. One suitable approach is the use of logarithmic amplifiers, which achieve a measurement range of up to 8 orders of magnitude (Figure 3), and another is dynamic signal amplification to adapt the output voltage to the current signal level [47]. Both approaches can and need to be closely integrated with the sensor element to achieve a high signal quality also under field conditions. Similarly, electronics suitable for field use were developed for Electrical Impedance Spectroscopy (EIS), allowing novel sensor self-monitoring strategies for MOX sensors [48,49] as well as for Gate Bias Cycling (GBC) of SiC-FETs [41] in both cases combined with TCO.

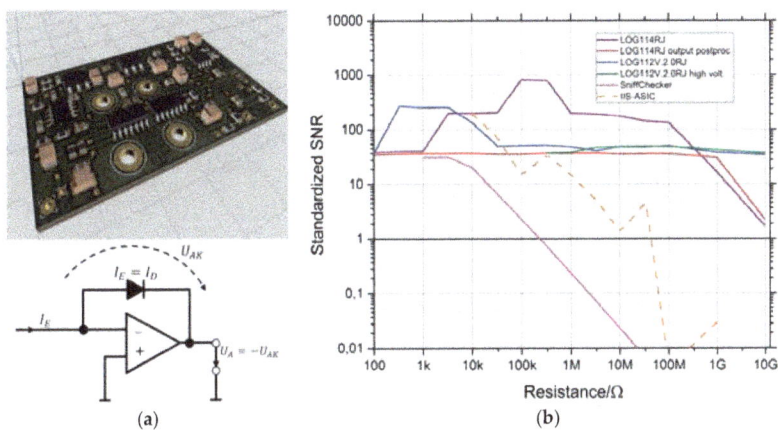

Figure 3. (**a**) Operating principle of a logarithmic amplifier based on the exponential current-voltage characteristic of a diode or transistor and PCB implementation. (**b**) Signal-to-Noise-Ratio (SNR) standardized to the same bandwidth of different logarithmic amplifiers compared to conventional electronics based on a linear AD converter ("SniffChecker" by 3S) and an integrated ASIC solution based on dynamic variation of measurement voltage and gain factor (IIS-ASIC, [47]).

4. Novel Integrated Pre-Concentrator Gas Sensor Microsystem

Despite the impressive sensitivities and very low detection limits that are achieved with MOS and SiC-FET sensors, some application targets are still difficult to achieve either due to very low target gas concentrations well below 1 ppb, e.g., trimethylamine with an odor threshold of 0.21 ppb; ethyl acrylate (and hydrogen sulfide) with an odor threshold of 0.47 ppb [50], or due to strong interference by other gases. Especially for measurements at very low gas concentrations, adsorbent materials are often used to achieve the required detection limits. Typical methods for detection of low VOC concentrations are based on sampling a defined gas volume using, e.g., Tenax® tubes. From these, the adsorbed gas is thermally desorbed during subsequent lab analysis based on gas chromatography (GC), often coupled with mass spectrometry (MS), to allow sensitive and selective gas detection. This approach has been miniaturized with the goal to achieve sensor systems for nearly continuous monitoring operating in adsorption/desorption cycles. However, these systems today are typically based on closed pre-concentrators combined with micro-pumps leading to systems that are not low-cost [51,52]. We have developed a new approach based on open pre-concentrators, i.e., absorbing material based on metal organic frameworks (MOF) deposited on a micro-hotplate similar to the gas sensor substrates.

These μ-pre-concentrators are integrated with the sensors in a common package with a small gas access. VOCs present in the ambient enter the package through the gas access and accumulate in the MOF material which has a large inner surface reaching partition coefficients orders of magnitude better than standard Tenax® [53]. Heating the μ-pre-concentrator will release the adsorbed gas molecules resulting in a considerably increased gas concentration within the sensor package, which is detected by the sensors [54]. Figure 4 illustrates the function principle of this novel approach which is based on gas transport by diffusion only thus requiring a miniaturized packaging solution. The function principle is also illustrated in a video to allow better understanding of the complex interaction within the microsystem [55]. Note that after release of the target gases from the pre-concentrator and subsequent cool-down, the pre-concentrator actually achieves a zero-air atmosphere within the package, at least for all gases adsorbing on the pre-concentrator [54]. This will actually allow an internal reference, as the sensors are briefly exposed only to permanent gases such as CO and H_2, thus improving the performance for target VOC detection and quantification by taking this into account in the signal evaluation. Furthermore, discrimination of different VOCs can be improved by taking into account the desorption temperature, i.e., for slow heating of the pre-concentrator. To make full use of this potential to boost sensitivity and selectivity, an application specific operating mode has to be designed as slow heating will decrease the peak concentration, while fast heating will release all VOCs simultaneously, thus limiting the selectivity. This novel integration approach is compatible with existing mass fabrication technologies and achieves sensor systems with greatly improved sensitivity at very low cost—at large volumes, the cost for the integrated system with, e.g., two sensors and one pre-concentrator could be less than one euro.

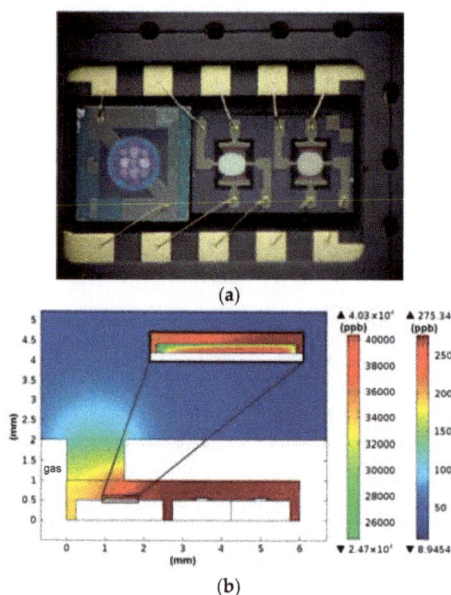

Figure 4. (a) Novel integrated pre-concentrator gas sensor microsystem combining a micro-pre-concentrator realized by deposition of MOF material on a micro-hotplate (left) with one or two gas sensors (right) in a single SMD package. Gas access is through a small opening above the pre-concentrator only. (b) Simulated gas concentrations inside the pre-concentrator material (left scale) and in the air (right scale), simulated for benzene and HKUST-1 as pre-concentrator material 1.5 s after start of desorption at 200 °C. The highest gas concentration is obtained inside the microsystem, i.e., at the locations of the two gas sensor chips S1 and S2. Adapted from [54].

5. Sensor System Testing and Evaluation

The impressive sensitivities and very low detection limits reported above can be achieved under well-defined laboratory conditions, which are, however, not easily achieved—in fact, very few test systems actually allow reliable testing of sensors at ppb and sub-ppb levels in a complex matrix. A pre-requisite for testing ultra-low concentrations is a suitable test setup, i.e., a gas mixing apparatus allowing exact control of gas admixtures under realistic and controlled ambient conditions. Some publications present very low detection limits, but these are sometimes just extrapolations from measurements at (much) higher concentrations or achieved in pure nitrogen as carrier gas, i.e., without oxygen or humidity, which has a huge influence on many sensors. Furthermore, standard zero air, which is also used for mixing of test gases, typically contains contaminations of approx. 10 ppm (zero air 5.0) and even the best zero air standards still contain approx. 1 ppm of unwanted and uncontrolled contaminations. While many of these do not influence the measurement (i.e., noble gases or CO_2), relevant trace contaminations can occur at concentrations orders of magnitude higher than the target gases to be tested. As the zero air used for the test gas is different from that used in the rest of the test setup, the observed sensor response can be caused by the contamination and not by the intended target gas when standard test gases with very low concentrations are used for direct measurement at ppb-level. Suitable approaches are the use of permeation tubes to introduce the target gas directly into the (zero air) carrier gas stream or two-stage gas dilution, allowing the use of test gases with concentrations much higher than any contaminations [56]. In both cases, the gas to be tested is injected into a carrier gas stream, which still contains contaminations, but the setup, which uses the same zero air throughout the system, ensures that the concentration of all contaminants stays constant, allowing the measurement of the sensor response to a change of only the target gas concentration. Further sources of error can be due to either VOC sources inside the system, e.g., from lubricants in valves or mass flow controllers, or due to adsorption of VOCs on inner surfaces preventing the target gas from reaching the sensor. The latter is especially problematic von SVOC which will then slowly diffuse out of the system, which can lead to large carry-over or "memory" effects. These effects can only by monitored by regular reference measurements of the overall system with analytical methods. Note that a gas test bench for broadband sensors such as MOS and SiC-FET should be based on a flow-through approach instead of gas re-circulation as reaction products from the interaction of the test gas with the sensor might otherwise lead to false results.

Finally, to accurately reflect measurements in ambient air, the typical mixture of natural air, which contains, in addition to nitrogen, oxygen, CO_2, and RH, approx. 1.8 ppm methane (CH_4), 550 ppb H_2, 325 ppb nitrous oxide (N_2O), and 150 ppb CO [57,58], has to be taken into account. CO shows the strongest variations with an annual cycle between 100 and 250 ppb [58]. While the effect of CH_4 is negligible, even the low-level exposure to H_2 and CO can easily change the baseline resistance of MOS sensors by one order of magnitude. At the same time, this will also reduce the sensitivity to other gases and distort the response pattern of sensor arrays and virtual multisensors. To achieve realistic test results, relevant background gases therefore have to be added to the zero air in gas test systems, and the natural variations have to be taken into account when determining detection limits and quantification resolution.

6. Factory and On-Site Calibration

To make full use of the potential of low-cost sensors for ultra-low VOC concentrations, calibration is an often underestimated challenge. The more complex the expectations, i.e., several target gases, mixtures of target gases, complex and variable background, the more complex the calibration procedure. This is due to the fact that data analysis is not based on a physical model of the sensor(s) but instead only on calibration data combined with pattern recognition techniques [29]—the calibration therefore has to span the full range of gases and concentrations expected in the later application. Due to slight variations in the individual sensors, at least part of this calibration has to be performed for each sensor individually, which can considerably contribute to the overall cost of the final gas sensor system. Furthermore, as the background can vary considerably for individual application environments, an extended calibration on-site might be necessary—this basically depends on the expected data quality to be achieved. While a scale-up of standard test benches (see above) can allow efficient factory calibration even for large production volumes, a test with various test gas cylinders would be impossible for on-site calibration. To address this issue, we have developed a novel approach for on-site calibration [59] based on VOCs dissolved in squalane, a long-chain alkane with low vapor pressure. This approach also allows producing a "zero air" atmosphere on-site within a confined volume in which the sensor and a vessel with squalane are kept: due to the large Henry constant, practically all VOCs present in the ambient will dissolve in the squalane resulting in a practically VOC free reference atmosphere (which will, however, still contain most inorganic gases such as CO, H_2, and NO_x). Note that this is similar to the pre-concentrator briefly achieving a VOC free atmosphere after cool-down, cf. Section 4. Similarly, using squalane loaded with a defined target VOC concentration will provide a source desorbing with a defined amount of the VOC, which can be used to calibrate sensors on-site. Note that this approach not only allows cost-efficient testing of the correct function of the sensor system, but also quantitative re-calibration to counteract sensor drift.

7. Conclusions and Outlook

Bringing together the different aspects outlined above—highly sensitive sensor elements; optimized dynamic operation (TCO, EIS, GBC) of the sensors; high-performance electronics for dynamic operation combined with advanced signal processing to achieve high sensitivity, selectivity and stability; low-cost pre-concentration to boost sensitivity and selectivity further; gas test bench for ppb and sub-ppb VOC concentrations, efficient factory, and on-site calibration—is a pre-requisite for the systematic development of low-cost sensor systems for VOC detection in various applications and for their validation in field tests. This integrated approach is at the core of the EU project SENSIndoor [60] addressing indoor air quality and demand controlled ventilation based on sensor systems placed in each room, i.e., offices, living and sleeping rooms, public buildings, transport, etc. On the one hand, the approach is used for optimization of the sensor elements themselves, i.e., based on novel nanotechnology approaches such as Pulsed Laser Deposition (PLD) and novel sensor materials [23,24,61–64]. On the other hand, extensive field tests are required which include reference tests based on existing standards. These developments will also lead to the development of new standards for VOC testing because existing standards do not cover the high spatial and temporal resolution that can be achieved with networks based on low-cost sensor systems. This aspect is currently addressed in the KEY-VOCs project under the European Metrology Research Program (EMRP) [65,66].

Note that in this contribution we have addressed sensor systems as a somewhat abstract concept, i.e., a device able to detect and quantify specific VOCs against a background of interfering gases. Not addressed here, but equally relevant to providing solutions for real world problems is a somewhat wider view that includes a structural model similar to the ISO-OSI model used in the field of communications. Systems based on low-cost semiconductor gas sensor principles will never achieve a universal performance independent of their application field, i.e., specific interferents and ambient conditions. Instead, application specific sensor systems are required—systems that would function well in the context for which they are designed, i.e., workplace safety monitoring or indoor air quality control, but that cannot be used to cover the full range of applications outlined in Figure 1. The structural model would cover the full range from expected benefits and application parameters at the top end down to the VOC atmosphere and the sampling method at the lower end [67] for realization of application specific sensor systems. By following this approach, it will be possible to address many different applications in environmental monitoring, indoor air quality, and industrial safety and health applications based on VOC monitoring, thus attaining a better understanding of VOC sources and effects to achieve a safer and healthier environment for all.

Acknowledgments: This work was in part performed within the SENSIndoor project. The SENSIndoor project is funded by the European Union's Seventh Framework Programme for research, technological development and demonstration under grant agreement No. 604311. Some foundations were laid within the MNT-ERA.net project VOC-IDS for which funding by the German Ministry for Education and Research (BMBF, funding codes 16SV5480K and 16SV5482) is gratefully acknowledged. The authors also acknowledge many useful discussions and suggestions from members of the COST action TD1105 EuNetAir and the EMRP project KEY-VOCs.

Author Contributions: Andreas Schütze and Tilman Sauerwald coordinated the research on trace VOC measurements at LMT together, Tilman Sauerwald designed most of the experiments and developed the model for MOS sensors in TCO mode together with Tobias Baur. Tobias Baur performed and evaluated the measurements for optimization of TCO (Figure 2). Martin Leidinger performed the simulations and corresponding experiments for the novel integrated sensor system (Figure 4). Wolfhard Reimringer designed the electronics that were used in most experiments. Ralf Jung designed the log-amp electronics supported by Tobias Baur and performed the SNR measurements within his Bachelor thesis (Figure 3). Thorsten Conrad contributed to the concept of the paper and especially to the chapter on sensor system calibration. Andreas Schütze wrote the paper.

Conflicts of Interest: The authors declare no conflicts of interest.

References

1. Jones, A.P. Indoor air quality and health. *Atmos. Environ.* **1999**, *33*, 4535–4564. [CrossRef]
2. Brinke, J.T.; Selvin, S.; Hodgson, A.T.; Fisk, W.J.; Mendell, M.J.; Koshland, C.P.; Daisey, J.M. Development of new volatile organic compound (VOC) exposure metrics and their relationship to sick building syndrome symptoms. *Indoor Air* **1998**, *8*, 140–152. [CrossRef]
3. Burge, P.S. Sick building syndrome. *Occup. Environ. Med.* **2004**, *61*, 185–190. [CrossRef] [PubMed]
4. World Health Organization. WHO Guidelines for Indoor Air Quality: Selected Pollutants, Geneva (2010). Available online: http://www.euro.who.int/__data/assets/pdf_file/0009/128169/e94535.pdf (accessed on 31 December 2016).
5. Guo, H.; Lee, S.C.; Chan, L.Y.; Li, W.M. Risk assessment of exposure to volatile organic compounds in different indoor environments. *Environ. Res.* **2004**, *94*, 57–66. [CrossRef]
6. European Parliament, Council of the European Union: Directive 2008/50/EC of the European Parliament and of the Council of 21 May 2008 on Ambient Air Quality and Cleaner Air for Europe. Available online: http://eur-lex.europa.eu/legal-content/EN/TXT/?uri=celex%3A32008L0050 (accessed on 31 December 2016).
7. TRGS 910: Technische Regeln für Gefahrstoffe, Risikobezogenes Maßnahmenkonzept für Tätigkeiten mit krebserzeugenden Gefahrstoffen, Bundesanstalt für Arbeitsschutz und Arbeitsmedizin, Ausschuss für Gefahrstoffe. 2015. Available online: http://www.baua.de/de/Themen-von-A-Z/Gefahrstoffe/TRGS/TRGS-910.html (accessed on 31 December 2016).
8. Reimann, P.; Schütze, A. Fire detection in coal mines based on semiconductor gas sensors. *Sens. Rev.* **2012**, *32*, 47–58. [CrossRef]
9. Pearce, T.C.; Schiffman, S.S.; Nagle, H.T.; Gardner, J.W. *Handbook of Machine Olfaction: Electronic Nose Technology*; Wiley: Weinheim, Germany, 2006.
10. European Collaborative Action Indoor Air Quality & Its Impact on Man: Sampling Strategies for Volatile Organic Compounds (VOCs) in Indoor Air. Available online: http://www.buildingecology.com/iaq/useful-publications/european-collaborative-action-on-urban-air-indoor-environment-and-human-exposure-reports-1/ (accessed on 31 December 2016).
11. ISO 16000–3:2011, Indoor Air—Part 3: Determination of Formaldehyde and Other Carbonyl Compounds in Indoor Air and Test Chamber Air—Active Sampling Method (2011). Available online: http://www.iso.org/iso/catalogue_detail.htm?csnumber=51812 (accessed on 31 December 2016).
12. Watson, N.; Davies, S.; Wevill, D. Air Monitoring: New Advances in Sampling and Detection. *Sci. World J.* **2011**, *11*, 2582–2598. [CrossRef] [PubMed]
13. Morrison, S.R. Semiconductor Gas Sensors. *Sens. Act.* **1982**, *2*, 329–341. [CrossRef]
14. Leidinger, M.; Sauerwald, T.; Reimringer, W.; Ventura, G.; Schütze, A. Selective detection of hazardous VOCs for indoor air quality applications using a virtual gas sensor array. *J. Sens. Sens. Syst.* **2014**, *3*, 253–263. [CrossRef]
15. Bastuck, M.; Bur, C.; Sauerwald, T.; Schütze, A. Quantification of volatile organic compounds in the ppb-range using partial least squares regression. In Proceedings of the SENSOR 2015—17th International Conference on Sensors and Measurement Technology, Nuremberg, Germany, 19–21 May 2015.

16. Kohl, D.; Kelleter, J.; Petig, H. Detection of Fires by Gas Sensors. In *Sensors Update*; WILEY-VCH: Weinheim, Germany, 2001; Volume 9, pp. 161–223.

17. Bârsan, M.; Hübner, N.; Weimar, U. Conduction mechanism in semiconducting metal oxide sensing films: Impact on transduction. In *Semiconductor Gas Sensors*, 1st ed.; Jaaniso, R., Tan, O.K., Eds.; Woodhead Publishing: Cambridge, UK, 2013; pp. 35–63.

18. Comini, E. Metal oxide nano-crystals for gas sensing. *Anal. Chim. Acta* **2006**, *568*, 28–40. [CrossRef] [PubMed]

19. Ramgir, N.; Datta, N.; Kaur, M.; Kailasaganapathi, S.; Debnath, A.K.; Aswal, D.K.; Gupta, S.K. Metal oxide nanowires for chemiresistive gas sensors: Issues, challenges and prospects. *Colloids Surf. A* **2013**, *439*, 101–116. [CrossRef]

20. Llobet, E. Gas sensors using carbon nanomaterials: A review. *Sens. Actuators B Chem.* **2013**, *179*, 32–45. [CrossRef]

21. Wang, Y.; Yeow, J.T. W. A Review of Carbon Nanotubes-Based Gas Sensors. *J. Sens.* **2009**. [CrossRef]

22. Wulan Septiani, N.L.; Yuliarto, B. Review—The Development of Gas Sensor Based on Carbon Nanotubes. *J. Electrochem. Soc.* **2016**, *163*. [CrossRef]

23. Huotari, J.; Kekkonen, V.; Haapalainen, T.; Leidinger, M.; Sauerwald, T.; Puustinen, J.; Liimatainen, J.; Lappalainen, J. Pulsed laser deposition of metal oxide nanostructures for highly sensitive gas sensor applications. *Sens. Actuators B Chem.* **2016**, *236*, 978–987. [CrossRef]

24. Leidinger, M.; Huotari, J.; Sauerwald, T.; Lappalainen, J.; Schütze, A. Selective detection of naphthalene with nanostructured WO₃ gas sensors prepared by pulsed laser deposition. *J. Sens. Sens. Syst.* **2016**, *5*, 147–156. [CrossRef]

25. Baur, T.; Schütze, A.; Sauerwald, T. Optimierung des temperaturzyklischen Betriebs von Halbleitergassensoren. *Technisches. Messen.* **2015**, *82*, 187–195. [CrossRef]

26. Sauerwald, T. Model based improvement of temperature cycled operation of tin oxide gas sensors. In Proceedings of the IX International Workshop on Semiconductor Gas Sensors, Zakopane, Poland, 13–16 December 2015.

27. Zhang, S.P.; Lei, T.; Li, D.; Zhang, G.Z.; Xie, C. UV light activation of TiO₂ for sensing formaldehyde: How to be sensitive, recovering fast, and humidity less sensitive. *Sens. Actuators B Chem.* **2014**, *202*, 964–970. [CrossRef]

28. Jin, H.; Haick, H. UV regulation of non-equilibrated electrochemical reaction for detecting aromatic volatile organic compounds. *Sens. Actuators B Chem.* **2016**, *237*, 30–40. [CrossRef]

29. Reimann, P.; Schütze, A. Sensor arrays, virtual multisensors, data fusion, and gas sensor data evaluation. In *Gas. Sensing Fundamentals*; Kohl, C.-D., Wagner, T., Eds.; Springer Series on Chemical Sensors and Biosensors, Volume 15; Springer: Berlin Heidelberg, Germany, 2014; pp. 67–107.

30. Clifford, P.K.; Tuma, D.T. Characteristics of semiconductor gas sensors II. transient response to temperature change. *Sens. Actuators* **1983**, *3*, 255–281. [CrossRef]

31. Lee, A.P.; Reedy, B.J. Temperature modulation in semiconductor gas sensing. *Sens. Actuators B Chem.* **1999**, *60*, 35–42. [CrossRef]

32. Gramm, A.; Schütze, A. High performance solvent vapor identification with a two sensor array using temperature cycling and pattern classification. *Sens. Actuators B Chem.* **2003**, *95*, 58–65. [CrossRef]

33. Bur, C.; Reimann, P.; Andersson, M.; Schütze, A.; Spetz, A.L. Increasing the Selectivity of Pt-Gate SiC Field Effect Gas Sensors by Dynamic Temperature Modulation. *IEEE Sens. J.* **2011**, *12*, 1906–1913. [CrossRef]

34. Fricke, T.; Sauerwald, T.; Schütze, A. Study of Pulsed Operating Mode of a Microstructured Pellistor to Optimize Sensitivity and Poisoning Resistance. In Proceedings of the IEEE Sensors Conference 2014, Valencia, Spain, 2–5 November 2014.

35. Ankara, Z.; Schütze, A. Low Power Virtual Sensor System based on a Micromachined Gas Sensor for Security Applications and Warning Systems. In Proceedings of the Eurosensors XXII Conference 2008, Dresden, Germany, 7–10 September 2008.

36. Schüler, M.; Sauerwald, T.; Schütze, A. A novel approach for detecting HMDSO poisoning of metal oxide gas sensors and improving their stability by temperature cycled operation. *J. Sens. Sens. Syst.* **2015**, *4*, 305–311. [CrossRef]

37. Bur, C.; Andersson, M.; Lloyd Spetz, A.; Schütze, A. Detecting Volatile Organic Compounds in the ppb Range with Gas Sensitive Platinum gate SiC-Field Effect Transistors. *IEEE Sens. J.* **2014**, *14*, 3221–3228. [CrossRef]

38. Weimar, U.; Göpel, W.A.C. Measurements on tin oxide sensors to improve selectivities and sensitivities. *Sens. Actuators B Chem.* **1995**, *26–27*, 13–18. [CrossRef]

39. Bastuck, M.; Bur, C.; Lloyd Spetz, A.; Andersson, M.; Schütze, A. Gas identification based on bias induced hysteresis of a gas-sensitive SiC field effect transistor. *J. Sens. Sens. Syst.* **2014**, *3*, 1–11. [CrossRef]

40. Reimann, P.; Dausend, A.; Darsch, S.; Schüler, M.; Schütze, A. Improving MOS Virtual Multisensor Systems by Combining Temperature Cycled Operation with Impedance Spectroscopy. In Proceedings of the ISOEN 2011, International Symposium on Olfaction and Electronic Nose, New York, NY, USA, 2–5 May 2011.

41. Bur, C.; Bastuck, M.; Lloyd Spetz, A.; Andersson, M.; Schütze, A. Selectivity enhancement of SiC-FET gas sensors by combining temperature and gate bias cycled operation using multivariate statistics. *Sens. Actuators B Chem.* **2014**, *193*, 931–940. [CrossRef]

42. Kammerer, T.; Ankara, Z.; Schütze, A. GaSTON—A versatile platform for intelligent gas detection systems and its application for fast discrimination of fuel vapors. In Proceedings of the Eurosensors XVII Conference 2003, Guimarães, Portugal, 22–24 September 2003.

43. Conrad, T.; Hiry, P.; Schütze, A. PuMaH—A temperature control and resistance read-out system for microstructured gas sensors based on PWM signals. In Proceedings of the IEEE Sensors Conference 2005, Irvine, CA, USA, 31 October–3 November 2005.

44. Conrad, T.; Fricke, T.; Reimann, P.; Schütze, A. A versatile platform for the efficient development of gas detection systems based on automatic device adaptation. In Proceedings of the Eurosensors XX Conference 2006, Göteborg, Sweden, 17–20 September 2006.

45. 3S-Toolbox. Available online: http://www.3s-ing.de/3s-technology/3s-toolbox/?L=1 (accessed on 30 December 2016).

46. Reimringer, W.; Rachel, T.; Conrad, T.; Schütze, A. Outdoor odor nuisance monitoring by combining advanced sensor systems and a citizens network. In Proceedings of the ISOEN 2015, 16th International Symposium on Olfaction and Electronic Noses, Dijon, France, 28 June–1 July 2015.

47. Stahl-Offergeld, M.; Hohe, H.P.; Jung, R.; Leidinger, M.; Schütze, A.; Sauerwald, T. Highly integrated sensor system for the detection of trace gases. In Proceedings of the IMCS'16–16th International Meeting on Chemical Sensors, Jeju Island, Korea, 10–14 July 2016.

48. Schüler, M.; Sauerwald, T.; Schütze, A. Metal oxide semiconductor gas sensor self-test using Fourier-based impedance spectroscopy. *J. Sens. Sens. Syst.* **2014**, *3*, 213–221. [CrossRef]

49. Sauerwald, T.; Schüler, M.; Schütze, A. *Erforschung einer Strategie und Entwicklung einer Messplattform zur Selbstüberwachung von Gasmesssystemen auf Basis von Halbleitergassensoren*; Bundesministerium für Wirtschaft und Energie: Berlin, Germany, 2015.

50. Leonardos, G.; Kendall, D.; Barnard, N. Odor Threshold Determinations of 53 Odorant Chemicals. *J. Air Pollut. Control. Assoc.* **1969**, *19*, 91–95. [CrossRef]

51. Akbar, M.; Restaino, M.; Agah, M. Chip-scale gas chromatography: From injection through detection. *Microsyst. Nanoeng.* **2015**, *1*, 15039. [CrossRef]

52. Zampolli, S.; Elmi, I.; Stürmann, J.; Nicoletti, S.; Dori, L.; Cardinali, G.C. Selectivity enhancement of metal oxide gas sensors using a micromachined gas chromatographic column. *Sens. Actuators B Chem.* **2005**, *105*, 400–406. [CrossRef]

53. Leidinger, M.; Rieger, M.; Sauerwald, T.; Nägele, M.; Hürttlen, J.; Schütze, A. Trace gas VOC detection using metal-organic frameworks micro pre-concentrators and semiconductor gas sensors. *Procedia Eng.* **2015**, *120*, 1042–1045. [CrossRef]

54. Leidinger, M.; Rieger, M.; Sauerwald, T.; Alépée, C.; Schütze, A. Integrated pre-concentrator gas sensor microsystem for ppb level benzene detection. *Sens. Actuators B Chem.* **2016**, *236*, 988–996. [CrossRef]

55. SENSIndoor clip. Available online: http://sensindoor.eu/film (accessed on 30 December 2016).

56. Helwig, N.; Schüler, M.; Bur, C.; Schütze, A.; Sauerwald, T. Gas mixing apparatus for automated gas sensor characterization. *Meas. Sci. Technol.* **2014**, *25*, 055903. [CrossRef]

57. OSS Foundation: Atmospheric Composition. Available online: ossfoundation.us/projects/environment/global-warming/atmospheric-composition (accessed on 30 December 2016).

58. Deutscher Wetterdienst (DWD): Composition of the Atmosphere–Trace Gases–Carbon Monoxide. Available online: www.dwd.de/EN/research/observing_atmosphere/composition_atmosphere/trace_gases/cont_nav/co_node.html (accessed on 30 December 2016).

59. Schultealbert, C.; Baur, T.; Schütze, A.; Böttcher, S.; Sauerwald, T. A novel approach towards calibrated measurement of trace gases using metal oxide semiconductor sensors. *Sens. Actuators B Chem.* **2017**, *239*, 390–396. [CrossRef]

60. Nanotechnology-Based Intelligent Multi-SENsor System with Selective Pre-Concentration for Indoor Air Quality Control, Website of the SENSIndoor Project. Available online: www.sensindoor.eu/ (accessed on 30 December 2016).

61. Lappalainen, J.; Huotari, J.; Leidinger, M.; Baur, T.; Alépée, C.; Komulainen, S.; Puustinen, J.; Schütze, A. Tailored Metal Oxide Nanoparticles, Agglomerates, and Nanotrees for Gas Sensor Applications. In Proceedings of the IX International Workshop on Semiconductor Gas Sensors, Zakopane, Poland, 13–16 December 2015.

62. Kekkonen, V.; Alépée, C.; Liimatainen, J.; Leidinger, M.; Schütze, A. Gas sensing characteristics of nanostructured metal oxide coatings produced by ultrashort pulsed laser deposition. In Proceedings of the Eurosensors XXIX Conference 2015, Freiburg, Germany, 6–9 September 2015.

63. Puglisi, D.; Eriksson, J.; Andersson, M.; Huotari, J.; Bastuck, M.; Bur, C.; Lappalainen, J.; Schütze, A.; Lloyd Spetz, A. Exploring the Gas Sensing Performance of Catalytic Metal/Metal Oxide 4H-SiC Field Effect Transistors. *Mater. Sci. Forum* **2016**, *858*, 997–1000. [CrossRef]

64. Puglisi, D.; Eriksson, J.; Bur, C.; Schütze, A.; Lloyd Spetz, A.; Andersson, M. Catalytic metal-gate field effect transistors based on SiC for indoor air quality control. *J. Sens. Sens. Syst.* **2015**, *4*, 1–8. [CrossRef]

65. Metrology for VOC Indicators in Air Pollution and Climate Change, Website of the KEY-VOCs Project. Available online: www.key-vocs.eu (accessed on 30 December 2016).

66. Spinelle, L.; Gerboles, M.; Kok, G.; Sauerwald, T. Sensitivity of VOC Sensors for Air Quality Monitoring within the EURAMET Key-VOC project. In Proceedings of the Fourth EuNetAir Scientific Meeting, Linköping, Sweden, 3–5 June 2015.

67. Reimringer, W.; Howes, J.; Conrad, T. Implementation of Complex Gas Sensor Systems: Ideas for a Structural Model. In Proceedings of the Sixth EuNetAir Scientific Meeting, Prague, Czech Republic, 5–7 October 2016.

environments

MDPI

Review

Currently Commercially Available Chemical Sensors Employed for Detection of Volatile Organic Compounds in Outdoor and Indoor Air

Bartosz Szulczyński and Jacek Gębicki *

Department of Chemical and Process Engineering, Chemical Faculty, Gdansk University of Technology,
11/12 G. Narutowicza Str., 80-233 Gdańsk, Poland; bartosz.szulczynski@pg.gda.pl
* Correspondence: jacek.gebicki@pg.gda.pl; Tel.: +48-58-347-2752

Academic Editors: Ki-Hyun Kim and Abderrahim Lakhouit
Received: 5 January 2017; Accepted: 1 March 2017; Published: 6 March 2017

Abstract: The paper presents principle of operation and design of the most popular chemical sensors for measurement of volatile organic compounds (VOCs) in outdoor and indoor air. It describes the sensors for evaluation of explosion risk including pellistors and IR-absorption sensors as well as the sensors for detection of toxic compounds such as electrochemical (amperometric), photoionization and semiconductor with solid electrolyte ones. Commercially available sensors for detection of VOCs and their metrological parameters—measurement range, limit of detection, measurement resolution, sensitivity and response time—were presented. Moreover, development trends and prospects of improvement of the metrological parameters of these sensors were highlighted.

Keywords: chemical sensor; measurement; metrological parameters; VOC

1. Introduction

Human activity contributes to generation of pollution, the amount and quality of which often exceed utilization abilities of the natural environment. It leads to loss of biosphere balance and frequently to unpredictable results. The anthropogenic sources of pollution can be divided into 4 basic groups:

- energetic connected with mining processes (mines, drawing shafts) and fuel combustion,
- industrial engulfing heavy industry (crude oil processing, metallurgy, cement plants, organic chemistry industry), production and application of solvents, food industry, pharmaceutic industry and so on,
- traffic, road transport (mainly cars), air and water transport,
- municipal, farms, houses, storage and utilization of solid waste and sewage (landfills, treatment plants) [1–4].

Volatile organic compounds (VOCs) constitute an important fraction of gaseous pollutants over urbanized areas, which originates from exhaust gases, evaporation of petroleum products and utilization of organic solvents [5–9]. VOCs take part in many photochemical reactions that yield harmful or even toxic products. Volatile organic compounds can also cause serious health problems as a number of them exhibit toxic, carcinogenic, mutagenic or neurotoxic properties. Moreover, many of them possess malodorous character that can contribute to deterioration of air quality [10–12]. What is more, VOCs are present at different concentration levels in indoor air. People who live in the climatic zone of the Central Europe spend most of their lifetime indoor. A statistic adult spends about 80% of their life at home, workplace, school, restaurant, cinema, means of public transport, etc. [13–16].

Elder people and children spent even more time indoor. Hence, even relatively low concentration of the pollutants present in indoor air can constitute health hazard due to long time of impact. The amount of pollutants present in indoor air increases due to different reasons [17–19]:

- construction of hermetic buildings (preventing energy loss), which do not provide enough air exchange,
- implementation of construction and finish materials with not fully identified properties,
- decreasing of height and volume of rooms.

The World Health Organization (WHO) recognized volatile organic compounds as the most important pollutants of indoor air. Harmless air is defined as the one, in which the total content of VOCs is lower than 100 $\mu g/m^3$. Among ca. 500 volatile compounds identified so far and present in indoor air only a few were proved pathogenic. Nevertheless, many of them are believed to contribute to such symptoms as: allergies, headaches, loss of concentration, drying and irritation of mucous membrane of nose, throat and eyes, etc. [20,21]. A set of such symptoms is named "Sick Building Syndrome". Indoor air quality depends on ambient air (outdoor one) and indoor emitters such as:

- construction materials,
- finish materials (paints, lacquers, wallpapers, floor covering, expanded polystyrene boards),
- burning processes, tobacco smoking,
- cleaning and preservation substances.

The largest group of emitted compounds are hydrocarbons including aromatic ones: benzene, toluene and xylenes. A significant contribution also originates from esters of short-chain organic acids, terpenes, formaldehyde and phenols. However, correct evaluation of outdoor and indoor air is not an easy task. Air is a quite complicated system subjected to changes even in a short period of time. Recently observed dynamic progress in analytical methods and analytical instruments is a basis for obtaining reliable information on indoor and outdoor air condition and quality. However, this progress leads to an increase in the cost of monitoring and air quality evaluation, which significantly limits their widespread application. Hence, alternative methods of acquisition of the information on air quality are being sought. Special attention is paid to sensor techniques [22–24]. According to a definition "chemical sensor" is a device, which changes chemical information from the environment into analytically useful signal. Numerous advantages of the chemical sensors include: low cost of manufacturing, simple design and possibility of miniaturization as well as relatively good metrological parameters such as sensitivity, selectivity, measurement range, linearity or response time. VOCs concentration level in air determines selection of the chemical sensors for their measurement. In the case of measurement of exhaust gases and indoor air at workplace, applicability of the chemical sensors is relatively high due to practical and economic advantages of these sensors.

The authors of this paper want to present up-to-date solutions available on the market as far as the chemical sensors for measurement of volatile organic compounds in outdoor and indoor air are concerned.

2. Characteristics of the Chemical Sensors for Detection of VOCs

In the case of commercially available chemical sensors for detection of volatile organic compounds one can distinguish two main approaches: the sensors for identification of explosion risk including thermal sensors (pellistors) and the infrared radiation absorption sensors. The latter group intended for detection of toxic gases belonging to VOCs includes electrochemical, semiconductor with solid electrolyte and PID-type sensors (photoionization detectors) [25,26]. Obviously, the main objective of the manufacturers is elaboration of a sensor with the best selective properties and with the low limit of quantification (LOQ). Figure 1 presents the concentration range of VOCs present in ambient air, indoor air at workplace and in exhaust gases. The figure also shows, which commercially available sensors are designated for detection and measurement of VOCs in ambient air, indoor air at workplace and in exhaust gases. It can be observed that the PID-type and electrochemical sensors are characterized by the lowest LOQ values. Design and operation principle of these sensors were described below.

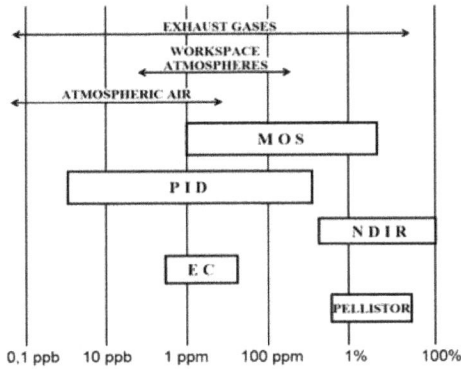

Figure 1. The concentration range of VOCs present in ambient air, indoor air at workplace and in exhaust gases and commercially available sensors are designed for detection and measurement of VOCs. MOS—Metal Oxide Semiconductor, PID—Photoionization Detectors, NDIR—Nondispersive infrared sensors, EC—Electrochemical, PELLISTOR—Thermal sensor.

2.1. Electrochemical (Amperometric) Sensors

In these sensors analyte particles diffuse through a membrane (separating gas environment from an internal electrolyte) and the internal electrolyte (most frequently aqueous solution of strong acids or bases, although mixtures with aprotic solvent are also utilized) towards the surface of a working electrode suitably polarized with respect to a reference electrode. Electrochemical reaction occurs at the working electrode, whereas a counter electrode experiences the reaction providing electron balance. A result of the redox reaction is generation of electric current being a sensor signal. This signal is proportional to concentration of the analyte present in direct vicinity of the sensor (gas environment) [27,28]. Figure 2 schematically illustrates design of the electrochemical sensor in a three-electrode version with the working (measurement) electrode, counter electrode and reference electrode of constant potential with respect to the working electrode.

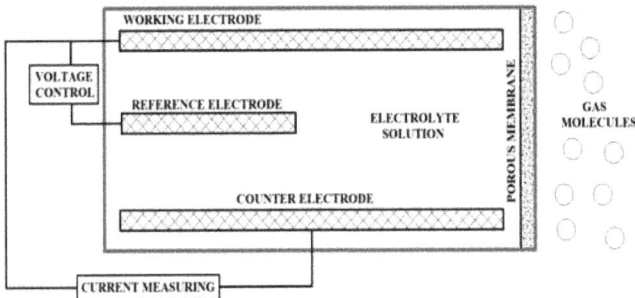

Figure 2. Scheme of electrochemical sensor in a three-electrode version with the measurement electrode, counter electrode and reference electrode.

Table 1 presents commercially available electrochemical (amperometric) sensors by Environmental Sensors Co. intended for measurement toxic compounds in air.

Table 1. Electrochemical sensors by Environmental Sensors Co. for measurement toxic compounds in air.

Pollutants	Range (ppm)	Resolution (ppm)	Response Time (s)
ammonia	0–50	0.5	150
carbon monoxide	0–1000	0.5	35
chlorine	0–20	0.1	60
ethylene oxide	0–20	0.1	140
formaldehyde	0–30	0.01	60
glutaraldehyde	0–20	0.01	60
hydrogen sulfide	0–50	0.1	30
nitric oxide	0–1000	0.5	10
nitrogen dioxide	0–20	0.1	35
sulfur dioxide	0–20	0.1	15

2.2. Metal Oxide Semiconductor Sensors

In these sensors analyte particles diffuse towards the receptor surface, which is metal oxide (maintained at suitable temperature using heater) where they undergo chemisorption. This interaction results in change of resistance of the receptor element. Two types of metal oxide semiconductors are utilized in measurement practice:

- type n (for example ZnO, SnO_2), which change resistance of the receptor element in the case of reducing gases presence,
- type p (for instance NiO, CoO), which change resistance of the receptor element in the case of oxidizing gases presence.

The sensing mechanism of semiconducting n-type metal oxides is based on a phenomenon of chemisorption of oxygen contained in air on the metal oxide layer. Adsorbed oxygen molecules trap the electrons from a conducting band of the semiconductor. It results in formation of energetic barriers between the grains of metal oxides, which block a flow of electrons. The consequence is an increase in resistance of the chemically sensitive layer of the sensor. Resistance of the chemically sensitive layer drops when gas molecules of reducing character appear. They react with bound oxygen leading to liberation of electrons. Reverse principle of operation occurs for p-type metal oxides, which identify oxidizing gases. The molecules of gas compounds remove electrons from the chemically sensitive layer, thus forming electron holes (charge carriers).

The process of signal generation (change of resistance) in the semiconductor sensors is not fully recognized. It is complex and consists of a number of co-existing phenomena: diffusion, chemisorption and desorption of gases, catalysed chemical reactions, electric conductivity of semiconductors and electron surface phenomena. The sensor sensitivity depends on:

- thickness of receptor layer and catalytic metal particles placed in it,
- temperature of receptor layer.

The receptor-converter elements of the resistance sensors can be:

- ultrathin (thickness from 5 to 100 nm)
- thin (thickness from 100 nm to 1 μm),
- thick (thickness from 1 to 300 μm).

This division has conventional character as it is not the lateral dimension but a method of elaboration that decides about classification into particular group. Desired measurement properties are obtained via surface modification of the receptor-converter element by:

- formation of ultrathin, discontinuous structures,
- deposition of thin layers having compact or microporous internal structure [29,30].

The materials used for the modification include metals: Pt, Pd, Ag, Au, V, Ru, Rh, Ti, Co, In and oxides: SiO_2, Rh_2O_3, RuO_2, Ir_2O_3, TiO_x, CuO, WO_3, V_2O_5.

The mechanism of operation of the semiconductor sensors strongly depends on temperature of the receptor-converter element as this parameter influences on the most important stages of a measurement process. Typical operation temperature of these sensors is within the 500–900 K range. Appropriate temperature is provided by electrical heaters.

The semiconductor sensors with solid electrolyte find a lot of applications in gas analysis. They can be used for measurement of hydrocarbons and their derivatives, alcohols, ethers, ketones, esters, carboxylic acids, nitroalkanes, amines or aromatic compounds [31–34]. Schematic design of the MOS-type sensor is shown in Figure 3.

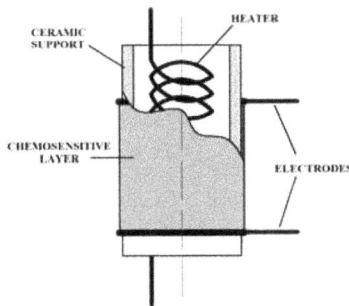

Figure 3. Scheme of MOS-type sensor.

2.3. Nondispersive Infrared Sensors (NDIR)

Flammable gases and vapours from the VOCs group are subjected to characteristic absorption of radiation from the infrared range. The ranges of oscillation frequency (wave number) characteristic for selected functional groups of VOCs are presented in Figure 4. The principle of operation of this type of sensor consists in arranging a source of infrared radiation along an optical line with a detector. When an analysed gas appears in a measurement chamber, it absorbs radiation of a particular wavelength and, following the Lambert-Beer law, there is a decrease in radiation reaching the detector, which is converted into electrical signal. Intensity of infrared radiation is diminished as it passes through the measurement cell. This reduction of light intensity is proportional to concentration of the gases or flammable vapours subjected to detection [35,36]. An important element of the sensor is an optical filter, which passes absorbed light of defined wavelength, thus providing selectivity of particular sensor. Some designs possess additional (reference) chamber, which is filled with non-absorbent gas (typically nitrogen). In this case the signal is generated based on a difference in readings from the detectors of both chambers. Figure 5 schematically presents a design of the NDIR-type sensor.

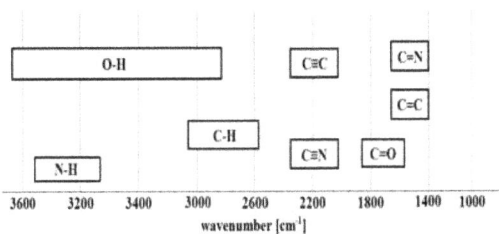

Figure 4. Ranges wave numbers characteristic for selected functional groups of volatile organic compounds.

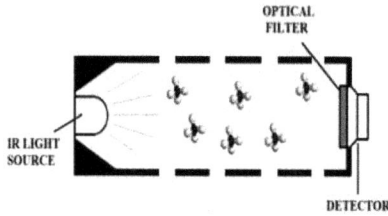

Figure 5. Scheme of NDIR-type sensor.

2.4. Thermal Sensor (Pellistor)

A phenomenon of explosion can be initiated in a mixture of flammable gas and air only within precisely defined concentration range. Lower explosion limit (LEL) determines the minimum concentration of the substance, which can react in a rapid combustion process. Upper explosion limit (UEL) describes the maximum amount of the fuel, at which the mixture contains enough oxidizer to initiate the explosion. The values of LEL and UEL differ for various substances and are usually expressed with respect to air. Concentrations of explosive substances below LEL and above UEL allow for safe operation. Table 2 presents the values of LEL for selected substances from the VOCs group [25]. The principle of operation of this type of sensor consists in diffusion of a mixture of air and particular flammable compound through porous sinter towards porous sensor surface. The porous element contains miniature coil made of platinum wire. Electric current flows through a coil made of platinum wire and heats the pellistor up to a few hundred degrees Celsius. The reaction at the catalytic surface releases heat, which increases temperature of the platinum coil, inducing an increase in its resistance. The pellistor is most commonly implemented as one arm of the Wheatstone bridge, the output of which is the final signal. In the case of temperature changes the output bridge signal is proportional to heat of reaction. Increase in temperature is a measure of concentration of flammable gas substance [25,37,38]. A scheme of the pellistor design is illustrated in Figure 6.

Table 2. Lower explosion limit for selected VOCs.

VOCs	Lower Explosion Limit (% *v/v*)	VOCs	Lower Explosion Limit (% *v/v*)
acetone	2.5	ethyl acetate	2.0
benzene	1.2	styrene	1.0
n-butanol	1.7	toluene	1.1
cyclohexane	1.0	1,3-butadiene	1.4
1,4-dioxane	1.9	n-butane	1.4
ethanol	3.1	methyl chloride	7.6
diethyl ether	1.7	dimethyl ether	2.7
methanol	6.0	ethylene oxide	2.6
n-hexane	1.0	methane	4.4
n-octane	0.8	propane	1.7

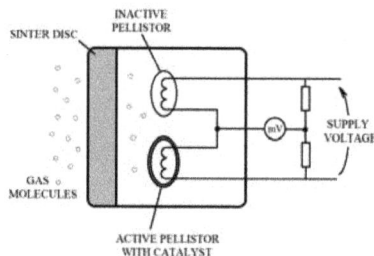

Figure 6. Scheme of pellistor-type sensor.

2.5. Photoionization Sensor (PID)

The principle of operation of the photoionization sensors consists in ionization (decay into charged particles) of neutral molecules of chemical compounds. When diffusing VOCs molecules enter the region of UV lamp impact, they are ionized by photons. Then formed ions are directed between two polarized electrodes. The ions move towards the electrodes in an electric field generated by an electrometer. In this way a current flow is generated, which is then converted into voltage signal. This signal is proportional to concentration of the compounds subjected to ionization. The photoionization sensors utilize electrodeless ultraviolet lamps (wavelength 10–400 nm). Operation of the lamp consists in excitation of the filling gas (most often krypton, xenon, radon) via the impact of external electromagnetic field. This type of sensor is most frequently applied for measurement of summary concentration of volatile organic compounds [39–41]. A scheme of the photoionization sensor is presented in Figure 7.

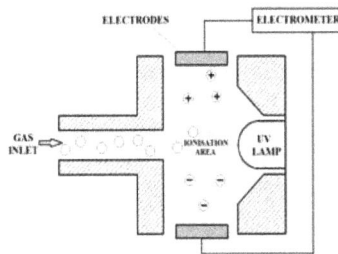

Figure 7. Scheme of photoionization sensor design.

Table 3 shows the main applications of the above described sensors for detecting the VOC.

Table 3. Main applications currently commercially available of sensors for determination of VOC.

Sensor Type	Applications	Compounds
MOS	- Urban air monitoring - Roadside monitoring - Industrial perimeter measurement - Indoor Air Quality - Smart home & Internet of Things modules - Medical equipment - Fire detection - Ventilation control/air cleaners	alcohols, aldehydes, aliphatic hydro-carbons, amines, aromatic hydro-carbons (petrol vapors, etc.), carbon oxides, CH4, LPG, ketones, organic acids.
PID	- Industrial hygiene & safety monitoring - Confined space entry - Soil contamination and remediation - Hazmat sites and spills - Arson investigation - Low concentration leak detection - EPA method 21 and emissions monitoring	VOC's with proper ionisation potential (isobutylene, aromatic hydrocarbons)
NDIR	- Indoor Air Quality - Combustion process monitoring - Biogas production	infrared absorbing VOC's (especially methane)
EC	- Breathanalyzer - Environmetnal protection - Odorants monitoring in natural gas applications - Urban and industrial area monitoring - Mobile monitoring applications	ethanol, formaldehyde, mercaptanes
PELLISTOR	- Hydrogen and combustible gas leak detectors - Detectors for fuel cells - Explosive atmosphere monitoring	most combustile gases and vapours (iso-butane, propane, methane)

The literature also provides information about other types of the sensors for VOCs detection in outdoor and indoor air. They include the following sensor types: chemoresistors (conductive polymers), surface acoustic wave, optical, quartz microbalance, FET, hybrid nanostructures. Table 4 presents advantages and disadvantages of these sensors together with the information about a VOC group they are designated.

Obviously, new solutions are still being sought in order to make the sensory techniques attractive and fulfil the customers' requirements. It is also worth to mention the sensor matrixes, also termed electronic noses, which are comprised of different types of chemical sensors. The most frequently utilized sensors include semiconductor, electrochemical and PID-type ones. These devices have been already applied in many fields of human activity including safety, environmental pollution, medicine, work safety regulations, food industry, chemical industry [42–44].

A search for new ways to improve metrological and utility parameters is also evident in the field of electrochemical and semiconductor sensors technology. In case of the semiconductor sensors with solid electrolyte it is proposed to modify the receptor layer via suitable doping and addition of catalysts. The conductometric sensors with a layer of organic conducting polymer operate in a way similar to the semiconductor sensors, change of resistance is a signal of both sensors. Two types of these sensors can be distinguished as far as their structure is concerned: with a composite polymer layer (for instance polypyrrole, graphite dissolved in a polymer matrix serving as an insulator) and with a layer of intrinsically conducting polymers (for example polyaniline, polypyrrole, polythiophene, polyacetylene, poly(phenylvinylene), poly(thienylenevinylene), poly(3,4-ethylenedioxythiophene), poly(N-vinylcarbazone). The sensors of the first type (utilizing the composite layer), as opposed to the MOS-type sensors, can operate at very high humidity and they exhibit linear response to many gas substances in a very wide concentration range. In case of the sensors of the second type (with the layer of intrinsically conducting polymers) the polymers can be doped (type n and type p) in a way similar to semiconductors. A type of doping results in an increase in the number of charge carriers and a change of polymer chain structure. Both processes cause enhanced mobility of holes or electrons in the polymer depending on the type of doping applied, which significantly contributes to an increase in sensor sensitivity. The advantages of the sensors with a layer of intrinsically conducting polymers include broad selection of suitable conducting polymer and possibility of its doping in order to obtain desired sensor's characteristics [45–50].

Traditional internal electrolyte in the electrochemical sensors is substituted with the ionic liquids, the main advantage of which is a possibility of modification of their physico-chemical properties by selection of suitable cation and anion. These modifications provide broadening of electrochemical durability range, relatively high electrical conductivity and possibility of measurement of volatile organic compounds, which would be impossible or significantly limited in the case of aqueous electrolytes utilization [51,52]. The ionic liquids also found application in conductive polymers sensors and optical sensors. In case of the conductive polymers sensors a function of the ionic liquids is to modify electrical conductivity of a polymer matrix and to provide selective sorption with respect to the air component determined. Suitable selection of the ionic liquid ensures improved sensitivity due to high enough gas/polymer matrix partition coefficient. Sensitivity of the optical sensors depends on a fluorescent dye (or dye mixture) applied and on the polymer matrix, in which the dye is dissolved. Such parameters of the matrix as polarity, hydrophobicity, porosity and expansion tendency have a significant impact on the value of sensor signal. The ionic liquids in the optical sensors influence on change of physico-chemical properties of the polymer matrix, mainly on polarity and sorption properties.

Table 4. Other chemical sensors used to detection VOC.

Sensor Type	Measurand	Advantages/Disadvantages	VOC's
Chemoresistors (conductive polymers)	conductivity	+ low operating temperature + small + low power consumption + cheap + sensitivity depends on the type of coating − sensitive to temperature and humidity − baseline drift (due to polymer instability) − short lifetime	acetone, acetonitrile, benzene, buthylamine, cyclohexane, ethanol, hexane, isopropanol, methanol, methylene chloride, toluene, xylenes [53–55]
Chemoresistors (graphene, carbon nanotubes composites)	conductivity	+ low detection limits + fast response + good sensitivity − complicated fabrication process − poor reproducibility	acetone, benzene, chloroform, ethanol, hexane, isopropanol, methanol, propanol, trichlotoethylene, toluene, m-xylene [56–60]
Hybrid nanostructures	depends on the type of sensor	+ low detection limits + high selectivity − complicated fabrication process − poor reproducibility	chloromethane, chloroform, ethanol, ethylacetate, methanol, octane, toluene [61–65]
Surface Acoustic Wave	frequency	+ small + low power consumption + god sensitivity to various chemicals + low detection limits − sensitive to humidity − large measurement noise − complicated signal processing system	ethanol, octane, toluene [66,67]
Optical	change in light parameters	+ portable and simple to use + possible visual detection + fast response time − Can be affected by humidity − very complex electronics − short lifetime due to photobleaching	benzene, butane, chlorobenzene, chloroform, dichloromethane, ethanol, ethyl acetate, formaldehyde, hexane, isopropanol, methane, methanol, oct-1-ene, propane, tetrahydrofurane, toluene, xylene [68–73]

Table 4. *Cont.*

Sensor Type	Measurand	Advantages/Disadvantages	VOC's
Quartz Microbalance	mass change	+ low detection limits + high sensitivity and selectivity + fast response − poor signal-to-noise performance − complicated signal processing system	acetone, acetonitrile, ethanol, 3-methyl-1-butanol, 1-octanol, toluene; p-xylene [74–76]
FET	threshold voltage change	+ low cost + small + reproducible − long response time − baseline drift − high working temperature − control of the surrounding environment	benzene, butylamine, ethanol, formaldehyde, hexane, hexanol, hexylamine, naphthalene, trimethylamine [77–80]
MEMS	depends on the type of sensor	+ small dimensions + on-chip integration + reproducible − complicated fabrication process − surface forces may dominate over other forces in the system − controlled working environment (dust-free) − development may be more costly	diethylamine, ethanol, isopropanol, ethane, methanol, propane, pentane, trimethylamine [81–83]

3. Commercially Available Chemical Sensors for Measurement of VOCs in Outdoor and Indoor Air

Table 5 gathers basic information about the types of commercially available sensors for measurement of toxic gases and flammable gasses from the VOCs group and their metrological parameters. The mentioned parameters include (unless the information was unavailable): measurement range, measurement accuracy, resolution, limit of detection, sensitivity and response time.

Table 5. Commercially available chemical sensors for measurement of volatile organic compounds.

Manufacturer	Sensor Type	Range	Accuracy	Resolution	LOD	Sensitivity	Response Time
Aeroqual	MOS	0–500 ppm	<±5 ppm + 10%	1 ppm	1 ppm	nd	30 s
	MOS	0–25 ppm	<±0.1 ppm + 10%	0.1 ppm	0.1ppm	nd	60 s
AppliedSensor	MOS	450–2000 ppm	nd	nd	nd	nd	nd
AMS	MOS	10–5000 ppm	nd	nd	10 ppm	0.002 (R_s/R_o)/ppm	<10 s
Cambridge CMOS Sensors	MOS	10–400 ppm	nd	nd	10 ppm	0.005 (R_s/R_o)/ppm	nd
SGX Snesortech	MOS	10–500 ppm	nd	nd	nd	0.014 (R_s/R_o)/ppm	nd
Mocon Baseline Series	PID	2–20,000 ppm	nd	nd	1–250 ppb	nd	<5 s
	PID	50 ppb–6000 ppm	nd	nd	1 ppb	nd	<3 s
Alphasense	PID	1 ppb–50 ppm	nd	<50 ppb	nd	nd	<3 s
	NDIR	0%–2.5%	<±500 ppm	<400 ppm	<500 ppm	nd	<40 s
Winsen	EC	0–1 mg/dm^3	nd	nd	nd	nd	<20 s
Winsen	EC	0–10 ppm	nd	0.02 ppm	nd	nd	<60 s
Citytech	EC	0–14 ppm	nd	<0.5 mg/m^3	nd	nd	<90 s
Figaro	MOS	1–100 ppm	nd	nd	nd	0.06 (R_s/R_o)/ppm	30 s
Umweltsensortechnik	MOS	0.1–100 ppm	<±20%	nd	nd	nd	<100 s
Wuhan Cubic	NDIR	0%–100%	±1% full scale	0.1%	nd	2%	<25 s
Unietc SRL	MOS	0.1–30 ppb	0.2 ppb	0.1 ppb	0.1 ppb	nd	nd
	EC	0.6–25 ppm	0.1 ppm	0.1 ppm	0.6 ppm	nd	nd
Synkera	MOS	50–900 ppm	±5% full scale	nd	50 ppm	nd	<60 s
ION Science	PID	0.1–6000 ppm, 1 ppb–40 ppm, 5 ppb–100 ppm	nd	nd	0.1 ppm, 1 ppb, 5 ppb	25 mV/ppm, 0.7 mV/ppm, 10 mV/ppm	3 s
Gray Wolf	PID	0.1–10000 ppm	nd	nd	nd	nd	<1 min
Environmental Sensors CO	EC	0–30 ppm	nd	0.01 ppm	0.1 ppm	nd	60 s
Z.B.P. SENSOR GAZ	pellistor	0%–100% LEL	±1.5% LEL	nd	nd	>30 mV/%	nd
Figaro	pellistor	0%–100% LEL	nd	nd	nd	0.02 mV/ppm	<30 s
SGX Snesortech	pellistor	0%–100% LEL	nd	nd	nd	15 mV/%	<10 s
MICROcel	pellistor	0%–100% LEL	nd	nd	nd	5 mV/%	<5 s
Sixth Sense	pellistor	0%–10% LEL	±10% LEL	nd	nd	>25 mV/%	<10 s

4. Conclusions

The methods of measurement of outdoor and indoor air pollutants belonging to the group of volatile organic compounds utilize broad spectrum of devices, from inexpensive chemical sensors presented in this paper to costly stationary systems such gas chromatographs, UV and IR spectrometers (including the ones with Fourier transformation), mass spectrometers, as well as electron capture detectors, flame ionization detectors, photoionization detectors and thermal conductivity detectors [84,85]. Selection of a suitable sensor depends on the gas to be measured, expected concentration range, the fact whether the sensor is meant to be stationary or portable, detect areal or point pollution, identify presence of other gases that could influence on the reading or damage the measurement device. Presented sensors are characterized by obvious advantages such as economic factor, relatively good metrological parameters, functionality, simple design, possibility of miniaturization. Nevertheless, they also possess certain limitations due to still too high limit of detection and quantification (MOS, NDIR, Pellistor, EC). Some of them also exhibit poor selective properties. That is why it is often the case that summary content of VOCs present in outdoor and indoor air is measured (PID). The chemical sensors reveal high reliability and functionality as far as personal, zone or indoor monitoring and identification of hazardous substances leaks from technological installations are concerned. This conclusion is also supported by the market, which offers wide variety of the chemical sensors for detection of flammable gases or toxic gases from the VOCs group. Progressively stricter legal rules concerning healthcare, protection of the natural environment and safety at workplace, focused on control of hazardous compounds emission beside industrial plants, power plants, transportation routes and municipal emitters such as landfills or sewage treatment plants resulted in the fact that air quality protection had become one of the most important elements of the ecological policy of the European Union. Despite the fact that the spectroscopic techniques are the most often applied reference methods for continuous ambient air monitoring (imision measurements), in many cases these are the chemical sensors, which are supplementary tools allowing prevention measures to be undertaken, especially in emission measurements.

Summarizing, the chemical sensors for detection and measurement of VOCs will be still developed and improved as there is an increasing market demand for them.

Acknowledgments: The investigations were financially supported by the Grant No. UMO-2015/19/B/ST4/02722 from the National Science Centre.

Author Contributions: Jacek Gębicki developed the concept and writing the manuscript. Bartosz Szulczyński made all the drawings and tables.

Conflicts of Interest: The authors declare no conflict of interest.

References

1. Lazarova, V.; Abed, B.; Markovska, G.; Dezenclos, T.; Amara, A. Control of odour nuisance in urban areas: The efficiency and social acceptance of the application of masking agents. *Water Sci. Technol.* **2013**, *68*, 614–621. [CrossRef] [PubMed]
2. Pearce, T.C.; Schiffman, S.S.; Nagle, H.T.; Gardner, J.W. *Handbook of Machine Olfaction*; WILEY-VCH Verlag GmbH & Co. KGaA: Weinheim, Germany, 2003.
3. Kampa, M.; Castanas, E. Human health effects of air pollution. *Environ. Pollut.* **2008**, *151*, 362–367. [CrossRef] [PubMed]
4. Gostelow, P.; Parsons, S.A.; Stuetz, R.M. Odour measurements for sewage treatment works. *Water Res.* **2001**, *35*, 579–597. [CrossRef]
5. Taylor, S.M.; Sider, D.; Hampson, C.; Taylor, S.J.; Wilson, K.; Walter, S.D.; Eyles, J.D. Community Health Effects of a Petroleum Refinery. *Ecosyst. Health* **2008**, *3*, 27–43. [CrossRef]
6. Henshaw, P.; Nicell, J.; Sikdar, A. Parameters for the assessment of odour impacts on communities. *Atmos. Environ.* **2006**, *40*, 1016–1029. [CrossRef]
7. Daud, N.M.; Sheikh Abdullah, S.R.; Abu Hasan, H.; Yaakob, Z. Production of biodiesel and its wastewater treatment technologies: A review. *Process Saf. Environ. Prot.* **2014**, *94*, 487–508. [CrossRef]
8. Yan, L.; Wang, Y.; Li, J.; Ma, H.; Liu, H.; Li, T.; Zhang, Y. Comparative study of different electrochemical methods for petroleum refinery wastewater treatment. *Desalination* **2014**, *341*, 87–93. [CrossRef]
9. Yavuz, Y.; Koparal, A.S.; Ogutveren, U.B. Treatment of petroleum refinery wastewater by electrochemical methods. *Desalination* **2010**, *258*, 201–205. [CrossRef]
10. Capelli, L.; Sironi, S.; Barczak, R.; Il Grande, M.; del Rosso, R. Validation of a method for odor sampling on solid area sources. *Water Sci. Technol.* **2012**, *66*, 1607–1613. [PubMed]
11. Bokowa, A.H. Review of odour legislation. *Chem. Eng. Trans.* **2010**, *23*, 31–36. [CrossRef]
12. Trincavelli, M.; Coradeschi, S.; Loutfi, A. Odour classification system for continuous monitoring applications. *Sens. Actuator B Chem.* **2009**, *139*, 265–273. [CrossRef]
13. Ilgen, E.; Karfich, N.; Levsen, K.; Angerer, J.; Schneider, P.; Heinrich, J.; Wichmann, H.E.; Dunemann, L.; Begerow, J. Aromatic hydrocarbons in the atmospheric environment: Part I. Indoor versus outdoor sources, the influence of traffic. *Atmos. Environ.* **2001**, *35*, 1235–1252. [CrossRef]
14. Chao, C.Y.H. Comparison between indoor and outdoor air contaminant levels in residential buildings from passive sampler study. *Build. Environ.* **2001**, *36*, 999–1007. [CrossRef]
15. Righi, E.; Aggazzotti, G.; Fantuzzi, G.; Ciccarese, V.; Predieri, G. Air quality and well-being perception in subjects attending university libraries in Modena (Italy). *Sci. Total Environ.* **2002**, *286*, 41–50.
16. Chan, A.T. Indoor–outdoor relationships of particulate matter and nitrogen oxides under different outdoor meteorological conditions. *Atmos. Environ.* **2002**, *36*, 1543–1551. [CrossRef]
17. Kot-Wasik, A.; Zabiegała, B.; Urbanowicz, M.; Dominiak, E.; Wasik, A.; Namieśnik, J. Advances in passive sampling in environmental studies. *Anal. Chim. Acta* **2007**, *602*, 141–163. [CrossRef] [PubMed]
18. Partyka, M.; Zabiegała, B.; Namieśnik, J.; Przyjazny, A. Application of Passive Samplers in Monitoring of Organic Constituents of Air. *Crit. Rev. Anal. Chem.* **2007**, *37*, 51–77. [CrossRef]

19. Weschler, C.J. Changes in indoor pollutants since the 1950s. *Atmos. Environ.* **2009**, *43*, 153–169. [CrossRef]
20. Zabiegała, B.; Partyka, M.; Zygmunt, B.; Namieśnik, J. Determination of volatile organic compounds in indoor air in the Gdansk area using permeation passive samplers. *Indoor Built Environ.* **2009**, *18*, 492–504. [CrossRef]
21. World Health Organization Publications. *Air Quality Guidelines for Europe*; European Series No. 91; World Health Organization: Copenhagen, Denmark, 2000.
22. Stetter, J.R.; Li, J. Amperometric gas sensors—A review. *Chem. Rev.* **2008**, *108*, 352–366. [CrossRef] [PubMed]
23. Rock, F.; Barsan, N.; Weimar, U. Electronic nose: Current status and future trends. *Chem. Rev.* **2008**, *108*, 705–725. [CrossRef] [PubMed]
24. Gebicki, J. Application of electrochemical sensors and sensor matrixes for measurement of odorous chemical compounds. *Trac Trends Anal. Chem.* **2016**, *77*, 1–13. [CrossRef]
25. Drager Technik fur das Leben, 2015. Available online: www.draeger.com (accessed on 15 August 2015).
26. Gebicki, J.; Dymerski, T. Application of Chemical Sensors and Sensor Matrixes to Air Quality Evaluation. In *The Quality of Air*, 1st ed.; de la Guardia, M., Armenta, S., Eds.; Elsevier: Amsterdam, The Netherlands, 2016; Volume 73, pp. 267–294.
27. Cao, Z.; Buttner, W.J.; Stetter, J.R. The properties and applications of amperometric gas sensors. *Electroanalysis* **1992**, *4*, 253–266. [CrossRef]
28. Bontempelli, G.; Comisso, N.; Toniolo, R.; Schiavon, G. Electroanalytical sensors for nonconducting media based on electrodes supported on perfluorinated ion-exchange membranes. *Electroanalysis* **1997**, *9*, 433–443. [CrossRef]
29. Chang, J.F.; Kuo, H.H.; Leu, I.C.; Hon, M.H. The effects of thickness and operation temperature on ZnO: Al thin film CO gas sensor. *Sens. Actuator B Chem.* **2002**, *84*, 258–264. [CrossRef]
30. Sakai, G.; Baik, N.S.; Miura, N.; Yamazoe, N. Gas sensing properties of tin oxide thin films fabricated from hydrothermally treated nanoparticles: Dependence of CO and H_2 response on film thickness. *Sens. Actuator B Chem.* **2001**, *77*, 116–121. [CrossRef]
31. Galdikas, A.; Mironas, A.; Setkus, A. Copper-doping level effect on sensitivity and selectivity of tin oxide thin-film gas sensor. *Sens. Actuator B Chem.* **1995**, *26*, 29–32. [CrossRef]
32. Yamazoe, N.; Sakai, G.; Shimanoe, K. Oxide semiconductor gas sensors. *Catal. Surv. Asia* **2003**, *7*, 63–75. [CrossRef]
33. Emelin, E.V.; Nikolaev, I.N. Sensitivity of MOS sensors to hydrogen, hydrogen sulfide, and nitrogen dioxide in different gas atmospheres. *Meas. Tech.* **2006**, *49*, 524–528. [CrossRef]
34. Berna, A. Metal Oxide Sensors for Electronic Noses and Their Application to Food Analysis. *Sensors* **2010**, *10*, 3882–3910. [CrossRef] [PubMed]
35. Arshak, K.; Moore, E.; Lyons, G.M.; Harris, J.; Clifford, S. A review of gas sensors employed in electronic nose applications. *Sens. Rev.* **2004**, *24*, 181–198. [CrossRef]
36. Munoz, R.; Sivret, E.C.; Parcsi, G.; Lebrero, R.; Wang, X.; Suffet, I.H.; Stuetz, R.M. Monitoring techniques for odour abatement assessment. *Water Res.* **2010**, *44*, 5129–5149. [CrossRef] [PubMed]
37. Brzózka, Z.; Wróblewski, W. *Sensory Chemiczne*; Oficyna Wydawnicza Politechniki Warszawskiej: Warsaw, Poland, 1998.
38. Wilson, A.D.; Baietto, M. Applications and advances in electronic-nose technologies. *Sensors* **2009**, *9*, 5099–5148. [PubMed]
39. Stetter, J.R.; Penrose, W.R. Understanding Chemical Sensors and Chemical Sensor Arrays (Electronic Noses): Past, Present, and Future. *Sens. Update* **2002**, *10*, 189–229. [CrossRef]
40. Wilson, A.D. Review of Electronic-nose Technologies and Algorithms to Detect Hazardous Chemicals in the Environment. *Procedia Technol.* **2012**, *1*, 453–463. [CrossRef]
41. Boeker, P. On "Electronic Nose" methodology. *Sens. Actuator B Chem.* **2014**, *204*, 2–17. [CrossRef]
42. Nicolas, J.; Romain, A.C. Establishing the limit of detection and the resolution limits of odorous sources in the environment for an array of metal oxide gas sensors. *Sens. Actuator B Chem.* **2004**, *99*, 384–392. [CrossRef]
43. Sohn, J.H.; Hudson, N.; Gallagher, E.; Dunlop, M.; Zeller, L.; Atzeni, M. Implementation of an electronic nose for continuous odour monitoring in a poultry shed. *Sens. Actuator B Chem.* **2008**, *133*, 60–69. [CrossRef]
44. Dentoni, L.; Capelli, L.; Sironi, S.; Rosso, R.; Zanetti, S.; Della Torre, M. Development of an Electronic Nose for Environmental Odour Monitoring. *Sensors* **2012**, *12*, 14363–14381. [CrossRef] [PubMed]

45. Albert, K.J.; Lewis, N.S.; Schauer, C.L.; Sotzing, G.A.; Stitzel, S.E.; Vaid, T.P.; Walt, D.R. Cross-Reactive Chemical Sensor Arrays. *Chem. Rev.* **2000**, *100*, 2595–2626. [CrossRef] [PubMed]
46. Munoz, B.C.; Steinthal, G.; Sunshine, S. Conductive polymer-carbon black composites-based sensor arrays for use in an electronic nose. *Sens. Rev.* **1999**, *19*, 300–305. [CrossRef]
47. Briglin, S.M.; Freund, M.S.; Tokumaru, P.; Lewis, N.S. Exploitation of spatiotemporal information and geometric optimization of signal/noise performance using arrays of carbon black-polymer composite vapor detectors. *Sens. Actuator B Chem.* **2002**, *82*, 54–74. [CrossRef]
48. Partridge, A.C.; Jansen, M.L.; Arnold, W.M. Conducting polymer-based sensors. *Mater. Sci. Eng. C* **2000**, *12*, 37–42. [CrossRef]
49. Bai, H.; Li, C.; Chen, F.; Shi, G. Aligned three-dimensional microstructures of conducting polymer composites. *Polymer* **2007**, *48*, 5259–5267. [CrossRef]
50. Bai, H.; Shi, G. Gas Sensors Based on Conducting Polymers. *Sensors* **2007**, *7*, 267–307.
51. Gebicki, J.; Kloskowski, A.; Chrzanowski, W.; Stepnowski, P.; Namiesnik, J. Application of Ionic Liquids in Amperometric Gas Sensors. *Crit. Rev. Anal. Chem.* **2016**, *46*, 122–138. [CrossRef] [PubMed]
52. Gebicki, J. Application of ionic liquids in electronic nose instruments. In *Analytical Applications of Ionic Liquids*; Koel, M., Ed.; World Scientific Publishing Europe Ltd.: London, UK, 2016; pp. 339–360.
53. English, J.T.; Bavana, A.D.; Freund, M.S. Biogenic amine vapour detection using poly (anilineboronic acid) films. *Sens. Actuator B Chem.* **2006**, *115*, 666–671. [CrossRef]
54. Li, B.; Santhanam, S.; Schultz, L.; Jeffries-EL, M.; Iovu, M.C.; Sauve, G.; Cooper, J.; Zhang, R.; Revelli, J.C.; Kusne, A.G.; et al. Inkjet printed chemical sensor array based on polythiophene conductive polymers. *Sens. Actuator B Chem.* **2007**, *123*, 651–660. [CrossRef]
55. Wang, F.; Yang, Y.; Swager, T.M. Molecular recognition for high selectivity in carbon nanotube/polythiophenechemiresistors. *Angew. Chem.* **2008**, *120*, 8522–8524. [CrossRef]
56. Lipatov, A.; Varezhnikov, A.; Wilson, P.; Sysoev, V.; Kolmakov, A.; Sinitskii, A. Highly selective gas sensor arrays based on thermally reduced grapheneoxide. *Nanoscale* **2013**, *5*, 5426–5434. [CrossRef] [PubMed]
57. Zito, C.A.; Perfecto, T.M.; Volanti, D.P. Impact of reduced graphene oxide on the ethanol sensing performance of hollow SnO_2 nanoparticles under humid atmosphere. *Sens. Actuator B Chem.* **2017**, *244*, 466–474. [CrossRef]
58. Tasaltin, C.; Basarir, F. Preparation of flexible VOC sensor based on carbon nanotubes and gold nanoparticles. *Sens. Actuator B Chem.* **2014**, *194*, 173–179. [CrossRef]
59. Castro, M.; Kumar, B.; Feller, J.F.; Haddi, Z.; Amari, A.; Bouchikhi, B. Novel e-nose for the discrimination of volatile organic biomarkers with an array of carbon nanotubes (CNT) conductive polymer nanocomposites (CPC) sensors. *Sens. Actuator B Chem.* **2011**, *159*, 213–219. [CrossRef]
60. Kumar, B.; Castro, M.; Feller, J.F. Poly(lactic acid)–multi-wall carbon nanotube conductive biopolymer nanocomposite vapour sensors. *Sens. Actuator B Chem.* **2012**, *161*, 621–628. [CrossRef]
61. Athawale, A.A.; Bhagwat, S.V.; Katre, P.P. Nanocomposite of Pd–polyaniline as a selective methanol sensor. *Sens. Actuator B Chem.* **2006**, *114*, 263–267. [CrossRef]
62. Santhanam, K.S.V.; Sangoi, R.; Fuller, L. A chemical sensor for chloromethanes using a nanocomposite of multiwalled carbon nanotubes with poly (3-methylthiophene). *Sens. Actuator B Chem.* **2005**, *106*, 766–771.
63. Sharma, S.; Nirkhe, C.; Prthkar, S.; Athawale, A.A. Chloroform vapour sensor based on copper/polyaniline nanocomposite. *Sens. Actuator B Chem.* **2002**, *85*, 131–136. [CrossRef]
64. Sayago, I.; Fernandez, M.J.; Fontecha, J.L.; Horrilli, M.C.; Vera, C.; Obieta, I.; Bustero, I. Surface acoustic wave gas sensors based on polyisobutylene and carbon nanotube composites. *Sens. Actuator B Chem.* **2011**, *156*, 1–5. [CrossRef]
65. Penza, M.; Antolini, F.; Antisari, M.V. Carbon nanotubes as SAW chemical sensors materials. *Sens. Actuator B Chem.* **2004**, *100*, 47–59. [CrossRef]
66. Sayago, I.; Fernandez, M.J.; Fontecha, J.L.; Horillo, M.C.; Vera, C.; Obieta, I.; Bustero, I. New sensitive layers for surface acoustic wave gas sensors based on polymer and carbon nanotube composites. *Sens. Actuator B Chem.* **2012**, *175*, 67–72. [CrossRef]
67. Viespe, C.; Grigoriu, C. Surface acoustic wave sensors with carbon nanotubes and SiO_2/Si nanoparticles based nanocomposites for VOC detection. *Sens. Actuator B Chem.* **2010**, *147*, 43–47. [CrossRef]
68. Crawford, M.; Stewart, G.; McGregor, G.; Gilchrist, J.R. Design of a portable optical sensor for methane gas detection. *Sens. Actuator B Chem.* **2006**, *113*, 830–836.

69. Goncalves, V.C.; Balogh, D.T. Optical chemical sensors using polythio-phene derivatives as active layer for detection of volatile organic compounds. *Sens. Actuator B Chem.* **2012**, *162*, 307–312. [CrossRef]

70. Elosua, C.; Arregui, F.J.; Zamarreño, C.R.; Bariain, C.; Luquin, A.; Laguna, M.; Mati, I.R. Volatile organic compounds optical fiber sensor based on lossy mode resonances. *Sens. Actuator B Chem.* **2012**, *173*, 523–529. [CrossRef]

71. Nizamidin, P.; Yimit, A.; Abdurrahman, A.; Itoh, K. Formaldehyde gas sensor based on silver-and-yttrium-co doped-lithium iron phosphate thin film optical waveguide. *Sens. Actuator B Chem.* **2013**, *176*, 460–466. [CrossRef]

72. Martínez-Hurtado, J.L.; Davidson, C.A.B.; Blyth, J.; Lowe, C.R. Holographic detection of hydrocarbon gases and other volatile organic compounds. *Langmuir* **2010**, *26*, 15694–15699. [CrossRef] [PubMed]

73. Wales, D.J.; Parker, R.M.; Quainoo, P.; Cooper, P.A.; Gates, J.C.; Grossel, M.C.; Smith, P.G.R. An integrated optical Bragg grating refractometer for volatile organic compound detection. *Sens. Actuator B Chem.* **2016**, *282*, 595–604. [CrossRef]

74. Khot, L.R.; Panigrahi, S.; Lin, D. Development and evaluation of piezoelectric-polymer thin film sensors for low concentration detection of volatile organic compounds related to food safety applications. *Sens. Actuator B Chem.* **2011**, *153*, 1–10. [CrossRef]

75. Si, P.; Mortensen, J.; Komolov, A.; Denborg, J.; Møller, P.J. Polymer coated quartz crystal microbalance sensors for detection of volatile organic compounds in gas mixtures. *Anal. Chim. Acta* **2007**, *597*, 223–230. [CrossRef] [PubMed]

76. Rizzo, S.; Sannicolo, F.; Benincori, T.; Schiavon, G.; Zecchin, S.; Zotti, G. Calix[4]arene-functionalized poly-cyclopenta[2,1-b;3,4-b]bithiophenes with good recognition ability and selectivity for small organic molecules for application in QCM-based sensors. *J. Mater. Chem.* **2004**, *14*, 1804–1811. [CrossRef]

77. Chang, J.B.; Liu, V.; Subramanian, V.; Sivula, K.; Luscombe, C.; Murphy, A.; Liu, J.; Fréchet, J.M.J. Printable polythiophene gas sensor array for low-cost electronic noses. *J. Appl. Phys.* **2006**, *100*, 14506–14507. [CrossRef]

78. Liao, F.; Yin, S.; Toney, M.F.; Subramanian, V. Physical discrimination of amine vapor mixtures using polythiophene gas sensor arrays. *Sens. Actuator B Chem.* **2010**, *150*, 254–263. [CrossRef]

79. Andersson, M.; Bastuck, M.; Huotari, L.; Lloyd Spetz, A.; Lappalainen, J.; Schütze, A.; Puglisi, D. SiC-FET Sensors for Selective and Quantitative Detection of VOCs Down to Ppb Level. *Procedia Eng.* **2016**, *168*, 216–220. [CrossRef]

80. Bur, C.; Bastuck, M.; Puglisi, D.; Schütze, A.; Lloyd Spetz, A.; Andersson, M. Discrimination and quantification of volatile organic compounds in the ppb-range with gas sensitive SiC-FETs using multivariate statistics. *Sens. Actuator B Chem.* **2015**, *514*, 225–233. [CrossRef]

81. Pandya, H.J.; Chandra, S.; Vyas, A.L. Integration of ZnO nanostructures with MEMS for ethanol sensor. *Sens. Actuator B Chem.* **2012**, *161*, 923–928. [CrossRef]

82. Pohle, R.; Weisbrod, E.; Hedler, H. Enhancement of MEMS-based Ga₂O₃ Gas Sensors by Surface Modifications. *Procedia Eng.* **2016**, *168*, 211–215. [CrossRef]

83. Kilinc, N.; Cakmak, O.; Kosemen, A.; Ermek, E.; Ozturk, S.; Yerli, Y.; Ozturk, Z.Z.; Urey, H. Fabrication of 1D ZnO nanostructures on MEMS cantilever for VOC sensor application. *Sens. Actuator B Chem.* **2014**, *202*, 357–364. [CrossRef]

84. Williams, M.L. Monitoring of exposure to air pollution. *Sci. Total. Environ.* **1995**, *168*, 169–174. [CrossRef]

85. Strang, C.R.; Levine, S.P.; Herget, W.F. A preliminary evaluation of the Fourier transform infrared (FTIR) spectrometer as a quantitative air monitor for semiconductor manufacturing process emissions. *Am. Ind. Hyg. Assoc. J.* **1989**, *50*, 70–77. [CrossRef]

environments

MDPI

Article

Indoor Air Quality Assessment and Study of Different VOC Contributions within a School in Taranto City, South of Italy

Annalisa Marzocca [1], **Alessia Di Gilio** [1,2], **Genoveffa Farella** [3], **Roberto Giua** [1] and **Gianluigi de Gennaro** [1,2,*]

[1] Regional Agency for Environmental Protection and Prevention of the Puglia (ARPA Puglia), Corso Trieste, 27, 70126 Bari, Italy; a.marzocca@arpa.puglia.it (A.M.); alessia.digilio@uniba.it (A.D.G.); r.giua@arpa.puglia.it (R.G.)
[2] Department of Biology University of Bari, via Orabona, 4, 70126 Bari, Italy
[3] Department of Chemistry, University of Bari, via Orabona, 4, 70126 Bari, Italy; jennyfarella@gmail.com
* Correspondence: gianluigi.degennaro@uniba.it or g.degennaro@arpa.puglia.it

Academic Editors: Ki-Hyun Kim and Abderrahim Lakhouit
Received: 31 December 2016; Accepted: 2 March 2017; Published: 10 March 2017

Abstract: Children spend a large amount of time in school environments and when Indoor Air Quality (IAQ) is poor, comfort, productivity and learning performances may be affected. The aim of the present study is to characterize IAQ in a primary school located in Taranto city (south of Italy). Because of the proximity of a large industrial complex to the urban settlement, this district is one of the areas identified as being at high environmental risk in Italy. The study carried out simultaneous monitoring of indoor and outdoor Volatile Organic Compounds (VOC) concentrations and assessed different pollutants' contributions on the IAQ of the investigated site. A screening study of VOC and determination of Benzene, Toluene, Ethylbenzene, Xylenes (BTEX), sampled with Radiello® diffusive samplers suitable for thermal desorption, were carried out in three classrooms, in the corridor and in the yard of the school building. Simultaneously, Total VOC (TVOC) concentration was measured by means of real-time monitoring, in order to study the activation of sources during the monitored days. The analysis results showed a prevalent indoor contribution for all VOC except for BTEX which presented similar concentrations in indoor and outdoor air. Among the determined VOC, Terpenes and 2-butohxyethanol were shown to be an indoor source, the latter being the indoor pollutant with the highest concentration.

Keywords: IAQ; indoor/outdoor ratio; real-time monitoring; industrial area

1. Introduction

Indoor pollution may have a significant bearing on health considering that people spend the majority of their time indoors, so understanding indoor exposures and the role of outdoor air pollution in shaping such exposures is a priority issue [1].

Among the indoor environments, school buildings are to be the object of attention, because children are more sensitive to pollutants than adults and they spend large amounts of time in the school environment [2–4]. Additionally, indoor air quality in school can significantly influence the efficiency of student learning processes and performances, determining an association between school absenteeism and poor building conditions [5–7].

Pollutants in school buildings can have different emission sources, operating both inside and outside the classroom. In these spaces, a number of factors influence good air quality including the number of inhabitants, activities conducted inside the classrooms and insufficient ventilation, aggravated by the poor construction and maintenance of many school buildings [8,9]. Among pollutants, volatile organic compounds (VOC) are considered important parameters for the assessment of air quality in indoor and outdoor environments because of their ubiquitous presence and their significant impact on the environment and human health [3,10].

A modest number of studies have examined VOC in school's indoor air [8,11–13]. Usually the organic compounds in these environments tend to be lower in the spacious and well ventilated classrooms with a low occupancy ratio [14], whereas accumulation of pollutants occurs when doors and windows remain closed for long periods of time in order to maintain thermal comfort, especially in winter. In addition to this, children are involved in different types of art and craft activities, using glue and paints that may increase the level of VOC in the classroom. Moreover, the use of cleaning products inside the school environment is also a possible source of VOC together with the intrusion of outdoor pollutants, especially if the school is located in an urban environment. The traffic, the distance from the road and the presence of industrial facilities in the neighborhoods, together with some meteorological factors (such as wind direction and speed), could affect the interior air quality. Moreover, atmospheric dispersion in the vicinity of buildings determines the concentration patterns near the buildings and on the building surfaces, and thus the infiltration of outdoor pollutants inside the buildings [15]. The ventilation rate in school classrooms has a large effect as it can lead to the intrusion of outdoor pollutants, especially in urban or industrial polluted areas. The Indoor-to-Outdoor concentration ratio (I/O) may be used to assess the relationship between indoor and outdoor VOC contributions.

The aim of the present study is the assessment of the indoor air quality in a primary school, located in close proximity to a high-impact industrial site such as the industrial area of Taranto (south of Italy), which is considered one of the areas of high environmental risk included in the list of polluted sites of national interest. In fact, the industrial zone of Taranto is characterized by a multiplicity of high-impact industrial activities, including the largest steel plant of Europe, a refinery, a quarry, a cement plant, a composting plant, a port in which liquid and solid bulk are handled, as well as containers.

The study carried out simultaneous monitoring of indoor and outdoor VOC concentration and assessed the influence of outdoor emissions on the IAQ of the investigated site. A screening study of VOC and determination of Benzene, Toluene, Ethylbenzene, Xylenes (BTEX) (VOC were carried out in three classrooms, in the corridor and in the yard of the school building. Simultaneously, total volatile organic compounds (TVOC) concentration was measured by means of real-time monitoring, in order to study the activation of sources during the monitored days. Data collected in indoor environments were compared with outdoor data to evaluate the different indoor/outdoor contributions and to study the intrusion phenomena.

2. Materials and Methods

The school selected is a naturally ventilated primary school located in a central position near three air quality stations (Orsini-Ilva network; Machiavelli network; Archimede-Regional network ARPA). Three classrooms were chosen on the basis of similar characteristics, such as level in the school building, surface, volume, number of windows, windows structure, number of occupants, activities, internal covering including flooring, wall and ceiling. Furthermore, measurements were carried out in the corridor of the school building and outdoors, in the yard of the school. Classrooms were occupied during school days for a total of 30 h during each monitored period (one week). They were unoccupied during afternoons, evenings, nights and Sunday. Class started at 08:00 and finished at 13:00.

A screening study of VOC and determination of BTEX, sampled with diffusive samplers suitable for thermal desorption, was carried out in the chosen sites. Two samplers were exposed for two monitoring periods (one week each) in all chosen environments. The indoor samplers were positioned at a height of about 1.5 m above the floor and at a distance that exceeded 1 m from windows or doors. Outdoor VOC measurements were collected for two weeks (two samplers for each monitoring period) at heights of about 2 m above the ground [16,17].

VOC monitoring was performed with Radiello® diffusive samplers (Fondazione Salvatore Maugeri, Padova, Italy) suitable for thermal desorption. The system consists of a cylindrical adsorbing cartridge made up of a stainless-steel net cylinder with 100 mesh, and an external diameter of 4.8 mm, containing 350 mg of Carbograph 4 (35–50 mesh). The cartridge was housed coaxially inside a cylindrical diffusive body of polycarbonate and microporous polyethylene. Before sampling, the cartridges were conditioned and analyzed to verify blank levels [18,19]. Each sampler was exposed for the periods indicated and then brought to the laboratory for analysis closed in a sealed glass tube. The desorption was carried out using a two-stage thermal desorber (Markes International Ltd., Unity™, Llantrisant, UK) equipped with an autosampler (mod. ULTRA™ TD, Markes International Inc., Cincinnati, OH, USA) which was provided with 100 positions and coupled with a gas chromatograph (Agilent GC-6890 PLUS, Loveland, CO, USA) and a mass selective detector (Agilent MS-5973N). In the first stage of desorption, the analytes were desorbed from the sample tube and refocused into a cold trap; then they were desorbed from the trap and carried into the gas chromatograph column [18–20].

Standard solutions were prepared by injecting successive dilution in methanol of a VOC standard mixture at 2000 µg/mL (Cus-5997 Ultra Scientific, Bologna, Italy). A calibration curve was prepared by injecting 1 µL of the standard solution into a tube; the spiked adsorbent tubes were then thermally desorbed in the same conditions of time, gas flow and split ratio as the samples. The sampling rates (Q), supplied by the manufacturer, were a function of the diffusive coefficient (D), which was the thermodynamic property of each chemical substance. To calculate the real concentration (C) of compounds in the atmosphere by GC quantification of analytes' mass (m), the sampling rates (Q) were used. When m was expressed in µg, the sampling period in minutes and C in µg/L, Q was expressed in L/min [21]. The assessment of the performance and reliability of the indoor monitoring methodology to determine VOC concentrations using radial diffusive samplers for thermal desorption was presented in previous works [18–20]. In particular, the repeatability of the analysis for thermal desorption, the complete desorption of the cartridges, the limit of detection (LOD), and the limits of quantification (LOQ) were evaluated [19]. The results showed that the RSD% was less than 10 for all compounds. The percentage recovery was higher than 95%, confirming the high method reliability for VOC analysis.

Simultaneously, real-time monitoring of TVOC was carried out in order to study the activation of sources during the monitored days and the possible intrusion of outdoor VOC in indoor air.

The high temporal resolution monitoring of TVOC was performed in the sampling sites with Corvus (Ion Science Ltd., Cambridge, UK) which uses Photo-ionization technology to detect a large range of VOC. Corvus was factory calibrated against isobutylene and thus the concentration of TVOC was equivalent to this gas.

3. Results and Discussion

The VOC concentrations measured (minimum, maximum and mean value), at all monitored sites during whole sampling period (two weeks), were presented in Table 1. Although the monitored school was located near a large industrial implant, in one of the district areas identified at high environmental risk in Italy, low outdoor concentrations were detected if compared with outdoor VOC concentrations found in similar industrial areas in the world [22–26]. These findings were probably due to the fact that the outdoor samplers were positioned in the yard of the school (according to UNI EN ISO 16000-1), so that detected concentrations were representative of the air in proximity of the indoor site. Outdoor sites were probably less affected by vehicular traffic and industrial emissions depending on the site-specific characteristics of the building and in particular of its yard. In fact, the court of the school is enclosed and has a little recess which isolated the entire school building from the closest industrial area.

Table 1. Volatile Organic Compounds concentrations measured in indoor and outdoor monitoring sites (minimum, maximum and mean value).

Compound	Classroom 5C			Classroom 5D			Classroom 4C			Corridor			Outdoor		
	Min	Max	Mean	Min	Max	Mean	Min	Max	Mean	Min	Max	Mean	Min	Max	Mean
Methyl-tert-butil- etere	0.33	0.82	0.53	0.28	0.86	0.61	0.14	0.80	0.47	0.32	0.68	0.50	0.65	0.96	0.85
Benzene	0.41	0.53	0.48	0.38	0.61	0.50	0.12	0.44	0.29	0.38	0.50	0.45	0.59	0.84	0.67
Toluene	1.34	1.78	1.54	1.51	1.88	1.66	1.37	1.49	1.45	1.31	1.69	1.52	1.54	1.90	1.64
N-Octane	0.21	0.56	0.40	0.11	1.25	0.61	0.23	0.36	0.26	0.21	0.35	0.28	0.20	0.35	0.26
Tetrachloroethylene	0.07	0.11	0.09	0.07	0.12	0.09	0.07	0.09	0.08	0.07	0.11	0.09	0.08	0.14	0.10
Ethylbenzene	0.48	1.18	0.80	0.41	1.40	0.76	0.43	1.18	0.71	0.42	1.13	0.77	0.37	1.23	0.76
M/p-Xylenes	0.85	1.74	1.27	0.79	1.64	1.21	0.89	2.20	1.44	0.81	1.45	1.14	0.78	1.54	1.14
Styrene	0.04	0.36	0.18	0.06	0.25	0.17	<0.03 *	0.43	0.17	0.04	0.23	0.13	0.13	0.17	0.15
O-Xylene	0.76	1.90	1.30	0.60	2.00	1.21	0.75	2.63	1.54	0.59	1.72	1.07	0.43	2.03	1.08
N-nonane	0.17	1.13	0.61	0.27	2.89	1.57	0.88	1.06	0.58	0.20	1.17	0.67	0.14	0.60	0.34
2-Butoxyethanol	27.92	35.55	31.60	33.32	46.43	36.31	23.10	36.59	30.21	28.61	33.36	30.30	<0.03 *	1.37	0.71
Alpha pinene	0.33	0.99	0.72	0.73	1.51	1.07	0.88	1.27	1.02	0.30	0.98	0.60	<0.03 *	0.15	0.06
Camphene	1.67	3.35	2.54	1.57	3.04	2.43	0.90	2.14	1.38	4.82	6.02	5.30	0.06	0.15	0.10
1,2,4-Trimethylbenzene	0.56	4.55	2.12	0.51	2.98	1.69	0.59	5.80	2.71	0.63	2.70	1.67	0.48	4.09	2.29
N-Decane	0.13	1.65	0.85	0.23	2.81	1.51	0.15	1.61	0.86	0.19	2.04	1.08	0.12	0.62	0.37
Limonene	1.14	3.75	2.19	1.33	4.53	2.94	3.65	7.98	5.66	1.65	3.56	2.57	<0.05 *	0.24	0.08

* Limit of detection (LOD).

In order to understand whether VOC sources were located in indoor or in outdoor air, the indoor concentration values for each pollutant were plotted against the corresponding outdoor value and five different I/O ranges were defined [8]. I/O ratios are commonly used to highlight the presence of important indoor emission sources. Ratios greater than a defined threshold value indicated the predominance of indoor contributions over outdoor contributions.

Figures 1–5 show that indoor concentrations of Benzene and substituted benzenes (during the first and the second monitoring campaign) are due to the input/intrusion of VOC from outdoor areas in most of the classrooms. Indeed, benzene and substituted benzenes are known as markers of vehicular traffic emissions [27].

Figure 1 shows that during both monitoring periods, all indoor sites present I/O ratios of Benzene in the range 0.5 < I/O < 2 except for the classroom 4C during the second week which presents an I/O ratio in the range 0.2< I/O <0.5. The indoor concentrations for BTEX were similar to outdoor ones; in fact, the I/O ratio value falls in the range 0.5 < I/O < 2. This is a demonstration that the I/O ratio does not provide sufficient values for highlighting critical issues, especially when values ranged from 0.5 to 2, in which case it is important to analyze the concentration levels. The I/O ratio can be used to highlight the presence of important indoor emission sources or outdoor emission sources when it is very high (I/O > 5) or very low (I/O < 0.2) respectively.

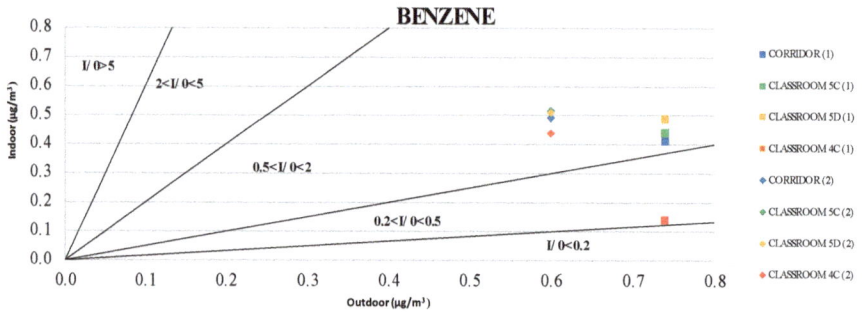

Figure 1. Benzene indoor concentrations against outdoor concentrations in all monitored sites during the first (1) and the second (2) monitoring periods.

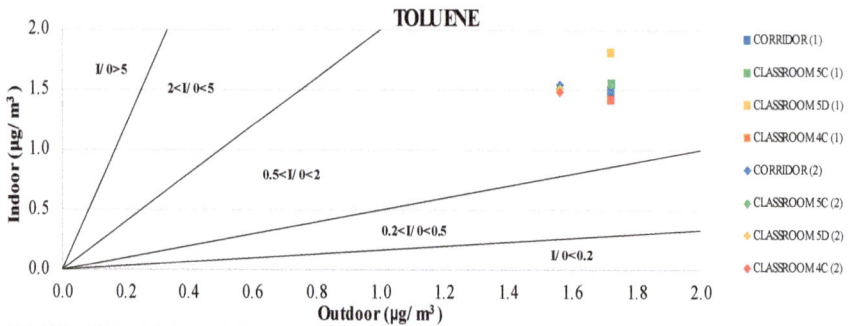

Figure 2. Toluene indoor concentrations against outdoor concentrations in all monitored sites during the first (1) and the second (2) monitoring periods.

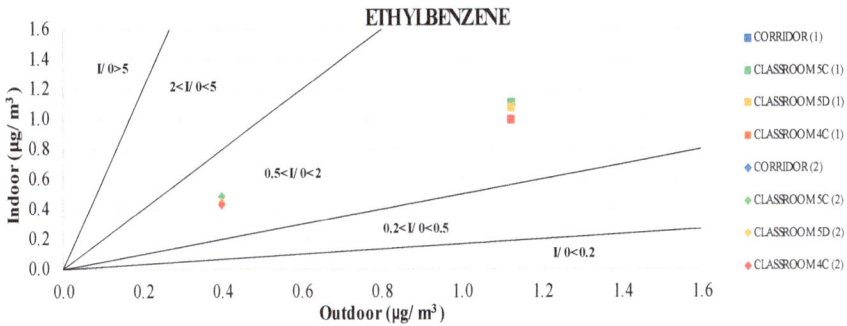

Figure 3. Ethylbenzene indoor concentrations against outdoor concentrations in all monitored sites during the first (1) and the second (2) monitoring periods.

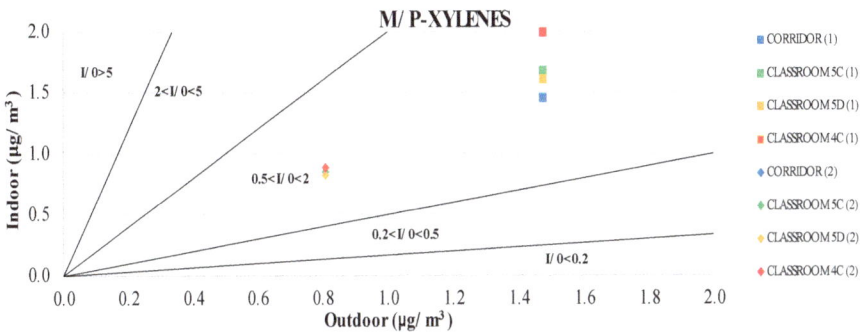

Figure 4. M, P, Xylenes indoor concentrations against outdoor concentrations in all monitored sites during the first (1) and the second (2) monitoring periods.

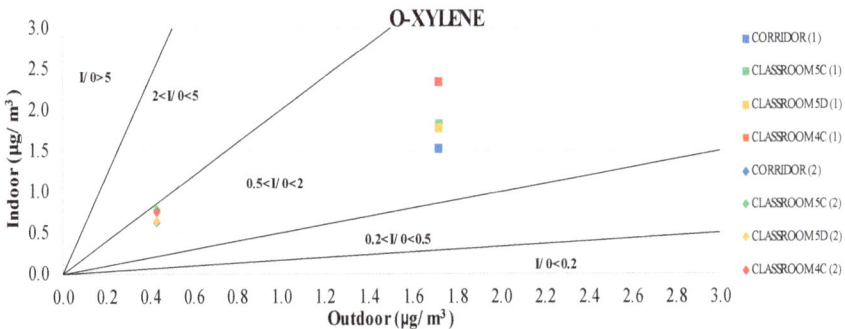

Figure 5. *O*-xylene indoor concentrations against outdoor concentrations in all monitored sites during the first (1) and the second (2) monitoring periods.

Figures 6 and 7 show indoor concentrations against outdoor concentrations for 2-butoxyethanol and limonene. For these pollutants, indoor average concentrations were respectively 30 and 5.9 times higher than outdoor ones.

2-Butoxyethanol, limonene and terpenes, in general, were the most abundant compounds in indoor air. Terpenes are odorous compounds mainly used to give pleasant fragrance in particular in cleaning products [8,28–30]. The high levels of 2-butoxyethanol in classrooms and in the corridor, were probably due to the fact that this compound is used as a solvent for many commercial products such as detergents, paints, adhesives, coatings, inks and products for personal care [31,32]. In particular, by reviewing the technical details of a range of cleaning products, it is present in percentages ranging between 2% and 20% in the degreaser and products for washing windows.

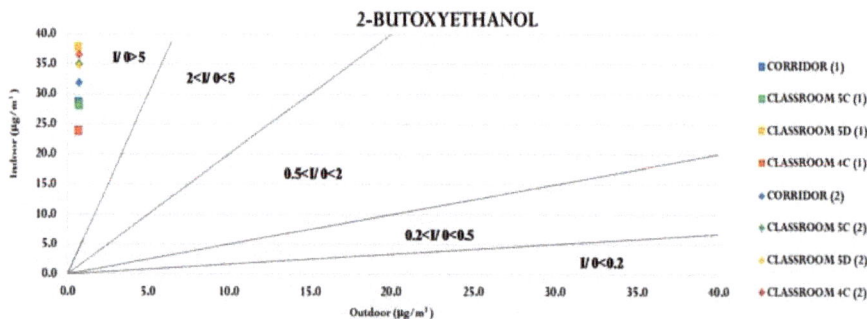

Figure 6. 2-Butoxyethanol indoor concentrations against outdoor concentrations in all monitored sites during the first (1) and the second (2) monitoring periods.

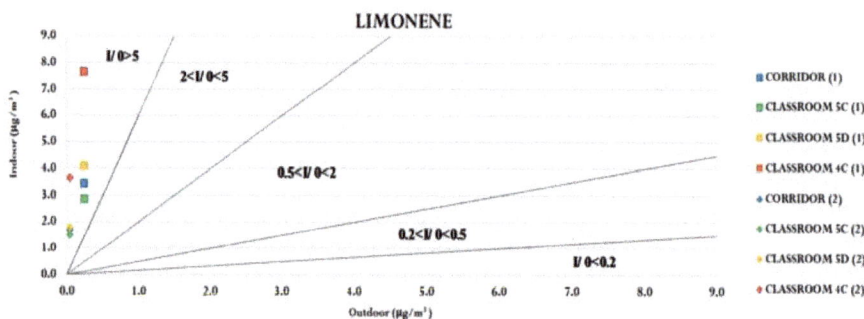

Figure 7. Limonene indoor concentrations against outdoor concentrations in all monitored sites during the first (1) and the second (2) monitoring periods.

In order to monitor the trend of TVOC concentration during school hours in the whole monitoring period, a high temporal resolution monitoring of Total Volatile Organic Compounds was also performed in the sampling sites. The detection of TVOC concentrations was carried out with Corvus (Ion Science Ltd., Cambridge, UK) which used Photo-ionization technology to detect a large range of VOC. This method can be useful to integrate the diffusive samplers' measurements value (a mean over the exposure periods—one week in this study) by identifying the trend of VOC concentration in different periods of the day over short time intervals—useful to find potential indoor sources.

Figure 8 shows the trend of TVOC concentrations during the sampling period.

The highest TVOC concentrations (ppm equivalent of isobutylene) in the monitored indoor sites were detected during cleaning activities, which occurred after the pupils leave and before the school closed (from 13:00). This confirms that cleaning activities were the most important source of indoor pollutants.

Table 2 shows a comparison of the indoor concentrations of selected VOC detected in this study with those obtained in studies conducted in other school microenvironments in other countries. As shown in the table, the VOC concentrations found at the school in Taranto City were in line with or above those of other studies conducted in the same condition (primary school monitored with diffusive samplers).

Table 2. Indoor school concentrations ($\mu g/m^3$) of selected VOC species in this, and in other studies. a = [11], b = [12], c = [13], d = [8].

Concentration $\mu g/m^3$

Compound	This Study		a		b (School 1)		b (School 2)		b (School 3)		c		d (School 1)		d (School 2)		d (School 3)		d (School 4)		d (School 5)		d (School 6)		d (School 7)		d (School 8)	
	In	Out	In	Out	In	Out	In	Out	In	Out	In	Out	In	Out	In	Out	In	Out	In	Out	In	Out	In	Out	In	Out	In	Out
Methyl-tert-butil-etere	0.52	0.85																										
Benzene	0.44	0.67	0.31	0.36	2.88	<LOD	3.01	3.13	2.54	2.46	0.09	0.06	0.59	0.55	1.27	0.62	3.11	0.44	0.32	1.00	0.60	0.81	0.77	0.60	0.18	0.37	0.08	0.14
Toluene	1.54	1.64	3.44	0.71	10.3	2	2.51	2.58	2.07	2.93	2.81	0.52	2.07	2.43	1.29	1.40	0.73	1.44	5.62	3.70	2.47	5.47	1.63		0.85	0.73	0.85	0.83
N-Octane	0.35	0.26																										
Tetrachloroethylene	0.09	0.10									0.02	<0.01	0.51	0.41	1.21	0.23	0.85	0.10	0.1	0.69	0.19	0.14	0.29	0.31	0.11	0.07	0.15	0.17
Ethylbenzene	0.76	0.76									0.24	<0.01	0.49	0.43	1.31	0.28	0.42	0.15	0.36	1.70	0.56	0.47	1.03	0.38	0.22	0.20	0.21	0.19
M/p-xylenes	1.22	1.14	0.32		8.8	1.22	1.42	1.26	2.82	1.78	2.3		1.47	1.46	1.83	0.89	10.81	0.41	0.65	2.38	1.80	1.65	3.75	1.13	0.65	0.57	0.58	0.60
Styrene	0.15	0.15									0.04	<0.01	0.73	0.45	1.24	0.21	0.17	0.14	0.29	1.26	0.31	0.18	0.46	0.26	0.18	0.16	0.13	0.12
O-Xylene	1.21	1.08			3.09	0.39	1.05	0.46	5.45	0.57	0.24	<0.01																
N-Nonane	0.80	0.34	0.8																									
2-Butoxyethanol	31.5	0.71																										
Alpha pinene	0.77	0.06	3.66		0.5	0.15			4.27	0.16	1.35	0.11	20.37	<0.03	4.97	2.89	1.36	0.27	4.99	3.08	1.07	0.23	2.85	<0.03*	2.59	0.34	1.33	0.72
Camphene	3.71	0.10											2.66	15.22	5.24	0.51	0.70	0.29	3.16	3.62	1.97	0.12	1.87	0.04	10.01	0.68	1.32	1.32
1,2,4-Trimethylbenzene	1.92	2.29									0.07																	
N-Decane	1.08	0.37			1	0.4	0.46	0.3	1.71	0.65			2.97	0.34	1.23	1.61	2.41	0.52	3.59	4.74	3.65	0.38	1.6	0.38	1.66	0.43	1.18	0.50
Limonene	3.08	0.08			3.17	0.39	1.71	0.65	86		4.41	1.2	18.27	32.15	10.10	10.25	4.15	1.03	7.34	10.51	3.26	0.36	3.52	0.08	5.16	1.22	1.96	1.01

* Limit of detection (LOD).

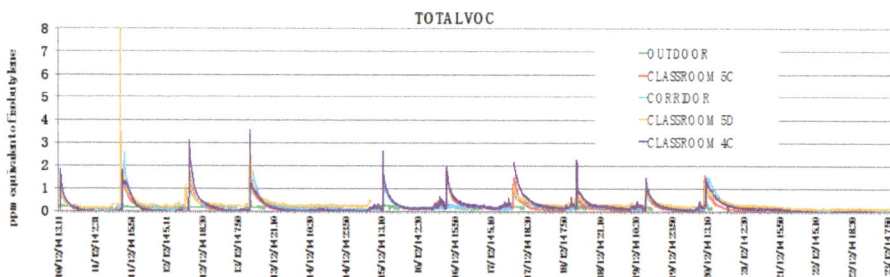

Figure 8. Trend of total volatile organic compounds (TVOC) concentration in all monitored sites for the entire sampling period.

4. Conclusions

The aim of this study was to assess the indoor air quality in a naturally ventilated school building by conducting a VOC screening monitoring campaign. The identification and quantification of VOC and the indoor/outdoor concentration plots allowed to detect the main emission sources. In particular, a significant indoor contribution by Terpenes and 2-butoxyethanol was found which presented the highest indoor concentrations. Despite the proximity of the school to the industrial area, the outdoor VOC concentrations were low in comparison to other studies conducted in the world in similar areas identified as being at high environmental risk. The use of high time resolution monitoring equipment facilitated the identification of the VOC emissions patterns of possible indoor sources and confirmed that the cleaning activities, occurring after the pupils leave, represents a critical issue for IAQ.

The present work represents a preliminary study and requires more measurements; in fact, analogous studies should be conducted in other schools, in particular when these are located near an industrial area, in order to give a precious tool for more efficient management with regards to mitigation actions.

Acknowledgments: Special appreciation goes to Dambruoso Paolo Rosario for the collaboration in sampling campaign. Annalisa Marzocca wrote the paper.

Author Contributions: This work is the result of the authors' commitment, starting from the idea and ending with its accomplishment. Particularly, each author contributed as follows: Annalisa Marzocca, Alessia Di Gilio, Gianluigi De Gennaro and Roberto Giua conceived and designed the sampling and the experiments; Genoveffa Farella analyzed the samples.

Conflicts of Interest: The authors declare no conflict of interest.

References

1. Ezzati, M. Indoor Air Pollution: Developing Countries. *Int. Encycl. Public Health* **2017**. [CrossRef]
2. Faustman, E.M.; Silbernagel, S.M.; Fenske, R.A.; Burbacher, T.M.; Ponce, R.A. Mechanisms underlying children's susceptibility to environmental toxicants. *Environ. Health. Perspect.* **2000**, *108*, 13–21. [CrossRef] [PubMed]
3. Dambruoso, P.R.; de Gennaro, G.; Demarinis Loiotile, A.; Di Gilio, A.; Giuncato, P.; Marzocca, A.; Mazzone, A.; Palmisani, J.; Porcelli, F.; Tutino, M. *School Air Quality: Pollutants, Monitoring and Toxicity*; Pollutant Diseases, Remediation and Recycling—Volume 4 of the Series Environmental Chemistry for a Sustainable World; Springer: Berlin, Germany, 2013; pp. 1–44.
4. Rivas, I.; Viana, M.; Moreno, T.; Pandolfi, M.; Amato, F.; Reche, C.; Bouso, L.; Àlvarez-Pedrerol, M.; Alastuey, A.; Sunyer, J.; et al. Child exposure to indoor and outdoor air pollutants in schools in Barcelona, Spain. *Environ. Int.* **2014**, *69*, 200–212. [CrossRef] [PubMed]
5. Wargocki, P.; Wyon, D.P. Ten questions concerning thermal and indoor air quality effects on the performance of office work and schoolwork. *Build. Environ.* **2017**, *112*, 359e366. [CrossRef]

6. Stafford, T.M. Indoor air quality and academic performance. *J. Environ. Econ. Manag.* **2015**, *70*, 34–50. [CrossRef]

7. Rosbach, J.T.M.; Vonk, M.; Duijm, F.; van Ginkel, J.T.; Gehring, U.; Brunekreef, B. A ventilation intervention study in classrooms to improve indoor air quality: The FRESH study. *Environ. Health* **2013**, *12*, 110. [CrossRef] [PubMed]

8. De Gennaro, G.; Farella, G.; Marzocca, A.; Mazzone, A.; Tutino, M. Indoor and Outdoor Monitoring of Volatile Organic Compounds in School Buildings: Indicators Based on Health Risk Assessment to Single out Critical Issues. *Int. J. Environ. Res. Public Health* **2013**, *10*, 6273–6291. [CrossRef] [PubMed]

9. Chithra, V.S.; Shiva Nagendra, S.M. Indoor air quality investigations in a naturally ventilated school building located close to an urban roadway in Chennai, India. *Build. Environ.* **2012**, *54*, 159–167. [CrossRef]

10. De Gennaro, G.; Dambruoso, P.R.; Demarinis Loiotile, A.; Di Gilio, A.; Giungato, P.; Marzocca, A.; Mazzone, A.; Palmisani, J.; Porcelli, F.; Tutino, M. Indoor air quality in schools. *Environ. Chem. Lett.* **2007**, *12*, 467–482. [CrossRef]

11. Pegas, P.N.; Nunes, T.; Alves, C.A.; Silva, J.R.; Vieira, S.L.A.; Caseiro, A.; Pio, C.A. Indoor and outdoor characterization of organic and inorganic compounds in city centre and suburban elementary schools of Aveiro, Portugal. *Atmos. Environ.* **2012**, *55*, 80e89. [CrossRef]

12. Pegas, P.N.; Evtyugina, M.G.; Alves, C.A.; Nunes, T.; Cerqueira, M.; Franchi, M.; Pio, C. Outdoor/indoor air quality in primary schools in Lisbon: A preliminary study. *Quim. Nova* **2010**, *33*, 1145–1149. [CrossRef]

13. Godwin, C.; Batterman, S. Indoor air quality in Michigan schools. *Indoor Air* **2007**, *17*, 109–121. [CrossRef] [PubMed]

14. Pegas, P.N.; Alves, C.A.; Evtyugina, M.G.; Nunes, T.; Cerqueira, M.; Franchi, M.; Pio, C.A.; Almeida, S.M.; Freitas, M.C. Indoor air quality in elementary schools of Lisbon in spring. *Environ. Geochem. Health* **2011**, *33*, 455e468. [CrossRef] [PubMed]

15. Santos, J.M.; Mavroidis, I.; Reis, N.C.; Pagel, E.C. Experimental investigation of outdoor and indoor mean concentrations and concentration fluctuations of pollutants. *Atmos. Environ.* **2011**, *45*, 6534–6545. [CrossRef]

16. ISO. *ISO 16000-1: Indoor Air—Part 1: General Aspects of Sampling Strategy*; ISO: Geneva, Switzerland, 2004.

17. ISO. *ISO 16000-5: Indoor Air—Part 5: Sampling Strategy for Volatile Organic Compounds (VOCs)*; ISO: Geneva, Switzerland, 2007.

18. Bruno, P.; Caputi, M.; Caselli, M.; de Gennaro, G.; de Rienzo, M. Reliability of a BTEX radial diffusive sampler for thermal desorption. *Atmos. Environ.* **2005**, *39*, 1347–1355. [CrossRef]

19. Bruno, P.; Caselli, M.; de Gennaro, G.; Iacobellis, S.; Tutino, M. Monitoring of volatile organic compounds in non-residential indoor environments. *Indoor Air* **2008**, *18*, 250–256. [CrossRef] [PubMed]

20. ISO. *ISO 16017-2: Indoor, Ambient and Workplace Air—Sampling and Analysis of Volatile Organic Compounds by Sorbent Tube/Thermal Desorption/Capillary Gas Chromatography—Part 2: Diffusive Sampling*; ISO: Geneva, Switzerland, 2003.

21. Radiello. Available online: http://www.radiello.com/english/index_en.htmlS (accessed on 10 December 2016).

22. Cetin, E.; Odabasi, M.; Seyfioglu, R. Ambient volatile organic compound (VOC) concentrations around a petrochemical complex and a petroleum refinery. *Sci. Total Environ.* **2003**, *312*, 103–112. [CrossRef]

23. Kalabokas, P.D.; Hatzaianestis, J.; Bartzis, J.G.; Papagiannakopoulos, P. Atmospheric concentrations of saturated and aromatic hydrocarbons around a Greek oil refinery. *Atmos. Environ.* **2001**, *35*, 2545–2555. [CrossRef]

24. Lin, T.Y.; Sree, U.; Tseng, S.H.; Hwa Chiu, K.; Wu, C.H.; Lo, J.G. Volatile organic compound concentrations in ambient air of Kaohsiung petroleum refinery in Taiwan. *Atmos. Environ.* **2004**, *38*, 4111–4122. [CrossRef]

25. Bruno, P.; Caselli, M.; de Gennaro, G.; de Gennaro, L.; Tutino, M. High spatial resolution monitoring of benzene and toluene in the Urban Area of Taranto (Italy). *J. Atmos. Chem.* **2006**, *54*, 177–187. [CrossRef]

26. Tiwari, V.; Hanai, Y.; and Masunaga, S. Ambient levels of volatile organic compounds in the vicinity of petrochemical industrial area of Yokohama, Japan. *Air Qual. Atmos. Health* **2010**, *3*, 65–75. [CrossRef] [PubMed]

27. Caselli, M.; de Gennaro, G.; Marzocca, A.; Trizio, L.; Tutino, M. Assessment of the impact of the vehicular traffic on BTEX concentration in ring roads in urban areas of Bari (Italy). *Chemosphere* **2010**, *81*, 306–311. [CrossRef] [PubMed]

28. Scorecard's Data Sources—Good Guide. Available online: http://scorecard.goodguide.com/about/txt/data.html (accessed on 9 December 2016).

29. Nazaroff, W.W.; Weschler, C.J. Cleaning products and air fresheners: Exposure to primary and secondary air pollutants. *Atmos. Environ.* **2004**, *38*, 2841–2865. [CrossRef]

30. De Gennaro, G.; Amodio, M.; Dambruoso, P.R.; de Gennaro, L.; Demarinis Loiotile, A.; Marzocca, A.; Stasi, F.; Trizio, L.; tutino, M. Indoor air quality (IAQ) assessment in a multistorey shopping mall by high-spatial-resolution monitoring of volatile organic compounds (VOC). *Environ. Sci. Pollut. Res. Int.* **2014**, *21*, 13186–13195.

31. Zhu, J.; Cao, X.L.; Beauchamp, R. Determination of 2-butoxyethanol emissions from selected consumer products and its application in assessment of inhalation exposure associated with cleaning tasks. *Environ. Int.* **2001**, *26*, 589–597. [CrossRef]

32. National Institute of Building Sciences. *Reviewing and Refocusing on IAQ in Schools*; National Institute of Building Sciences: Washington, DC, USA, 2006.

environments

MDPI

Article

Lean VOC-Air Mixtures Catalytic Treatment: Cost-Benefit Analysis of Competing Technologies

Gabriele Baldissone *, Micaela Demichela and Davide Fissore

Dipartimento di Scienza Applicata e Tecnologia, Politecnico di Torino, Corso Duca degli Abruzzi 24, 10129 Torino, Italy; micaela.demichela@polito.it (M.D.); davide.fissore@polito.it (D.F.)
* Correspondence: gabriele.baldissone@polito.it; Tel.: +39-011-090-4629

Received: 4 May 2017; Accepted: 18 June 2017; Published: 25 June 2017

Abstract: Various processing routes are available for the treatment of lean VOC-air mixtures, and a cost-benefit analysis is the tool we propose to identify the most suitable technology. Two systems have been compared in this paper, namely a "traditional" plant, with a catalytic fixed-bed reactor with a heat exchanger for heat recovery purposes, and a "non-traditional" plant, with a catalytic reverse-flow reactor, where regenerative heat recovery may be achieved thanks to the periodical reversal of the flow direction. To be useful for decisions-making, the cost-benefit analysis must be coupled to the reliability, or availability, analysis of the plant. Integrated Dynamic Decision Analysis is used for this purpose as it allows obtaining the full set of possible sequences of events that could result in plant unavailability, and, for each of them, the probability of occurrence is calculated. Benefits are thus expressed in terms of out-of-services times, that have to be minimized, while the costs are expressed in terms of extra-cost for maintenance activities and recovery actions. These variable costs must be considered together with the capital (fixed) cost required for building the plant. Results evidenced the pros and cons of the two plants. The "traditional" plant ensures a higher continuity of services, but also higher operational costs. The reverse-flow reactor-based plant exhibits lower operational costs, but a higher number of protection levels are needed to obtain a similar level of out-of-service. The quantification of risks and benefits allows the stakeholders to deal with a complete picture of the behavior of the plants, fostering a more effective decision-making process. With reference to the case under study and the relevant operational conditions, the regenerative system was demonstrated to be more suitable to treat lean mixtures: in terms of time losses following potential failures the two technologies are comparable (Fixed bed plant: 0.35 h/year and Reverse flow plant: 0.56 h/year), while in terms of operational costs, despite its higher complexity, the regenerative system shows lower costs (1200 €/year).

Keywords: cost-benefit analysis; VOC treatment; lean mixtures; reverse-flow reactor; Integrated Dynamic Decision Analysis

1. Introduction

Different technologies are available for the treatment of gaseous streams containing VOCs (Volatile Organic Compounds), namely catalytic or homogeneous combustion, absorption, adsorption, etc. The goal is either to recover the VOCs, or to destroy them, thus avoiding, in both cases, their emission into the atmosphere.

The treatment of "lean" streams, where the concentration of the VOCs is particularly low (e.g., lower than 1%, v/v), is a particularly challenging case study as the low concentration makes the VOC recovery technically and economically impractical. In this case, the catalytic combustion stands out as the reference technology as it allows fulfilling the constraints on the characteristics of the product released into the atmosphere [1]. Thus, in case of lean streams to be treated two competitive technologies can be regarded as effective: a catalytic fixed-bed reactor with a heat exchanger for heat recovery purposes, and an intensified plant, with a catalytic reverse-flow reactor, where regenerative heat recovery is achieved through the periodical reversal of the flow direction.

In fact, the temperature at which it is required to carry out the catalytic combustion ranges (in most cases) from 200 °C to 500 °C, depending on the characteristics of the catalyst used and on the chemical compounds that have to be removed, while the temperature of the gaseous stream to be treated can be significantly lower, in some cases close to room temperature. This poses the problem of energy recovery: a large amount of energy is required to pre-heat the feed to the reaction temperature, and a fraction of this energy can be recovered from the treated gas to pre-heat the feed. In any case, a certain amount of energy has to be supplied, as the efficiency of this approach is usually not higher than about 70% [2].

In this framework, a particularly effective technology is represented by the reverse-flow reactor, firstly proposed by Cottrell [3]. The operating principle of the reverse-flow reactor is particularly simple: at first, the reactor is pre-heated to the target temperature and, then, the gaseous stream is fed to the reactor. As a consequence, the gas is heated by the solid, thus reaching the reaction temperature (and cooling the solid material close to the entrance of the reactor), and, then, it heats the solid before leaving the reactor. After a certain time, which has to be carefully selected, the flow direction is reversed and, now, the hot gas, before leaving the reactor, heats the solid that has been cooled in the previous stage of reactor operation. In this way, a regenerative heat recovery is achieved in the reactor, where the ending sections act as two heat exchangers, storing the heat from the hot gas after combustion, and, then, pre-heating the cold gas fed to the reactor. As the ending sections of the reactor acts as heat exchangers, the catalyst here can be replaced by inert material, thus reducing catalyst cost [4]. This system was widely investigated in the past, both from the theoretical and the experimental points of view [5–9], evidencing the possibility of operating in presence of streams with composition and concentration variable in time [10,11], and also with a monolith type catalyst, to reduce the pressure drop in the system [12]. All these studies evidenced the key role of the switching time: a high switching frequency is required to enhance the "heat trap" effect due to the periodical flow reversal and, thus, to get autothermal combustion when feeding a very lean gas. Unfortunately, after each flow reversal a certain amount of gas and VOCs remaining in the entrance of the reactor is released into the atmosphere (wash-out phenomenon), and this is particularly relevant in case of high switching frequencies. Although some alternative reactor configurations have been proposed to cope with this problem (see, among the others, Kolios et al. [13]; Luzi et al. [14]), the use of an electrical heater in the central part of the system appears to be a particularly simple and effective solution to treat a low concentration stream, avoiding too high switching frequencies [2].

The above described defines the major aim of this paper: defining an optimized approach to support the decision-making process among possible plant and process alternatives, based on the risk associated with the different competing options. In particular, among the possible tools available for the risk-based decision making, the Integrated Dynamic Decision Analysis (IDDA) has been chosen, since it was demonstrated in several papers [15–18] to be more effective in assessing the risks, giving a more complete representation of the system behavior and also in facilitating the knowledge transfer. The risk considered for comparison purposes has been expressed in terms of costs, to be in line with the typical decisional process of the stakeholders and decision makers, thus adopting a cost-benefit analysis (CBA) approach.

As discussed in Saarikoski et al. [19], the cost-benefit analysis (CBA) is an economic evaluation method for comparing the costs and benefits of different projects or policy options [20]. CBA aims to value all impacts of project alternatives in monetary units, possibly discounted to a specified year, making it possible to screen or rank alternatives by a single monetary measure, often the net present value (NPV). The basic steps of CBA, according to Boardman et al. [21], can be summarized as follows:

1. The definition of the project options to be evaluated: in the present case, the two competing technologies for VOC treatment above described;

2. The decision on which costs and benefits are accounted for and the selection of the measurement method to evaluate all the costs and benefits: as described later on, the costs here used to compare the two alternatives are the operational costs, the variable costs to be taken into account with the investment costs for the equipment, and the time losses in case of malfunction;

3. The estimation of costs and benefits over a relevant time period and their conversion into a common currency, discounted into present value: dealing with operational costs, the reference period has been assessed over one year of operation.

4. Drawing the recommendations based on the costs and benefits and the sensitivity analysis, when available: the decision making was in this case supported not only by the costs figures, but also on the logical-probabilistic model of the two systems, jointly analyzed with the phenomenological model of the process through the Integrated Dynamic Decisional Analysis methodology.

The CBA is used in several field as: in the waste management [22], in the medical resource allocation [23], and in the energy management [24]. A sensitivity analysis was not performed, since all the input parameters required for plant design have been selected equal to the optimal values, according to the available know-how on these technologies, as it will be discussed in the following.

The Integrated Dynamic Decisional Analysis (IDDA) was firstly proposed by Galvagni [25,26] and, then, it was applied to different case studies, e.g., tank overflowing [15,16], risk-based design of an allyl-chloride production plant [27], analysis and optimization of procedures for LPG tanks maintenance and testing [28]. The IDDA methodology is based on the combined use of a logical-probabilistic model of the system under analysis and its phenomenological model. The logical-probabilistic model allows identifying all potential sequences of events the system can undergo and their probability of occurrence, and it can be coupled to a phenomenological model, i.e., a mathematical description of the process plant behavior both in case of normal condition and in case of equipment faults. The phenomenological model is used to evaluate the consequences for each sequence of events, and the overall risk value. The joint modelling, that still constitutes a novelty in the process domain of application, allows a risk-based decision making to be performed, as discussed in Piccinini & Demichela [15] and in Demichela & Camuncoli [16].

This paper is structured as follows: Chapter 2 is dedicated to the description of the two competing technologies under study, with the logical and probabilistic model and the phenomenological one developed to simulate and foreseen their behavior in case of correct functioning and in case of failures. Chapter 3 shows the results of the analysis, with a comparison of the performances of two plant alternatives in terms of costs and benefits. In the end, some technical and methodological conclusions are drawn.

2. Case Study

The case study considered in this paper is the treatment of 5000 Nm^3/h of a gaseous stream obtained from a polymerization plant. The stream is composed of ethylene glycol, 0.2% v/v, and nitrogen: oxygen has thus to be added to this gas as the VOC is removed through catalytic combustion. The gas temperature is 230 °C, while the pressure is 1.17 bar. After VOC removal (the target concentration of VOC in the product stream has to be lower than 50 ppm), the gaseous stream is recycled to the polymerization plant: as the presence of oxygen is highly undesired in the recycled stream, great care has to be paid when adding the oxygen in the feed, and when carrying out the catalytic combustion, in such a way that (almost) no unreacted oxygen is present in the product (the constraint considered in this study is that the oxygen concentration in the treated has to be lower than 0.01%).

The performance of two plants has been compared in this paper. The first is based on a traditional fixed-bed reactor, while the second comprises a reverse-flow reactor. In both cases the VOC is removed through catalytic combustion. In the fixed-bed reactor a commercial catalyst based on Platinum (0.15 w/w) and Palladium (0.15% w/w) over alumina (type K-02120 by Heraeus, Hanau, Germany) has been considered. The catalyst is supplied as pellets, with a diameter ranging from 2 to 4 mm, with a BET surface area of 115 m^2/g. The GHSV (Gas Hourly Space Velocity) suggested by the catalyst supplier is 10,000 h^{-1}, with an operating temperature of 350 °C. In the second reactor a metal oxides based commercial catalyst (EnviCat® VOC-1544 by Sud-Chemie, Munich, Germany), composed of cooper (3.34% w/w) and manganese (5.44% w/w) oxides over alumina, has been considered. The catalyst is supplied as pellets, with a diameter ranging from 4 to 6 mm, with a BET surface area of 73 m^2/g. The GHSV (Gas Hourly Space Velocity) suggested by the catalyst supplier is 15,000 h^{-1}, with an operating temperature of 430 °C. The rational for the choice of two different catalysts is the fact that in the reverse-flow reactor it is much easier to get higher temperatures (as it is just required to increase the switching frequency, without introducing additional heat) with respect to the fixed-bed reactor and, thus, the use of the metal oxides based catalyst, whose operating temperature is lower, allows reducing the catalyst cost.

For the reactor sizing, more conservative values of GHSV, with respect to those suggested by the catalyst suppliers, have been considered, namely 5000 h^{-1} for the Pt-Pd based catalyst and 10,000 h^{-1} for the Cu-Mn oxides based catalyst. Moreover, for the reverse-flow reactor based plant the use of inert particles, alumina spheres with the same diameter of the catalyst particles, has been considered to increase the heat capacity of the system: the same amount of inert and of catalyst has been assumed, and the reactor has been divided into three sections, namely a first inert layer, the catalyst layer and the second inert layer. By this way in both cases the reactor has the same size, namely 1.5 m diameter, and 0.6 m length.

A simplified sketch of the fixed-bed based plant is shown in Figure 1. Beside the catalytic reactor, the plant is composed of the following pieces of equipment:

- a gas blower (B01), used to feed the plant;
- an oxygen feeding line, used to feed the required amount of oxygen in the system, on the basis of the measurement of the oxygen concentration in the gas leaving the plant;
- a filter (F01), used to stop particles suspended in the gaseous stream;
- a heat exchanger (H01), used to pre-heat the feed using the hot gas leaving the reactor;
- a heater (H02), used to heat the feed to the combustion temperature,
- a reactor (R01), used to conduct the combustion reaction.

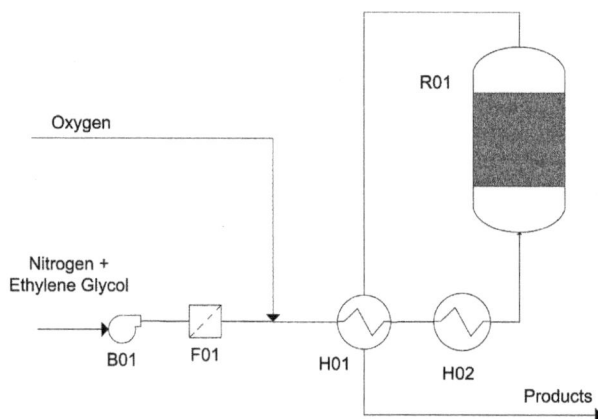

Figure 1. Sketch of the plant for the treatment of air-VOC mixtures in a steady-state fixed-bed reactor.

With respect to the alarms and control/protection systems used in this plant, the following can be listed:

- High concentration alarm in the feed line, indicating when the VOC concentration of the feed exceeds a certain threshold (0.5% in this case), as the high concentration of the feed would be responsible for a too high temperature in the reactor, with a consequent catalyst deactivation;

- High temperature alarm after the blower: it alerts the operator when the temperature is higher than a certain threshold (270 °C in this case) to prevent filter damages;
- Differential pressure alarm in the filter, indicating filter blocking;
- High/Low temperature alarms in the gas line exiting H01: if the temperature is higher than a certain threshold (450 °C in this case), then catalyst overheating and damages can occur, while if the temperature is lower than a certain threshold (300 °C in this case), then it would not be possible to reach the temperature required to get a full VOC removal in the reactor;
- High temperature alarm in the gas line exiting H02: it alerts the operator if the temperature is higher than a certain threshold (450 °C in this case) and, thus, catalyst damages can occur. Due to the importance of this alarm, temperature is measured by a series of sensors operating logic on voting (2:3);
- High temperature alarms in the reactor: when the temperature is higher than a first threshold (400 °C in this case) the operator is alerted and, if the temperature continues increasing, after a second threshold (450 °C in this case) a protection system is activated, causing the shut-down on the system;
- High oxygen concentration alarm in the product stream, indicating that too much oxygen is present in the product stream and, thus, it cannot be recycled to the polymerization plant;
- High VOC concentration alarm in the product stream, indicating that the VOC concentration is too high and, thus, that the catalytic reactor is not working properly.

A simplified sketch of the reverse-flow reactor based plant is shown in Figure 2. Besides some elements similar to the fixed-bed based plant, namely the blower B01, the filter F01, the reactor R01, and the oxygen feeding line, the plant comprises the following pieces of equipment:

- a heat exchanger (H01), used in the start-up phase and for control purposes (according to the design of Fissore et al. [9]);
- a set of valves (V01, V02, V03, V04) used to periodically reverse the gas flow direction: when valves V01 and V04 are open, and V02 and V03 are closed, the gas flows from the top to the bottom of the reactor, while the opposite occurs when valves V01 and V04 are closed and V02 and V03 are open. Valve V05 is usually closed, and used only for control purposes.

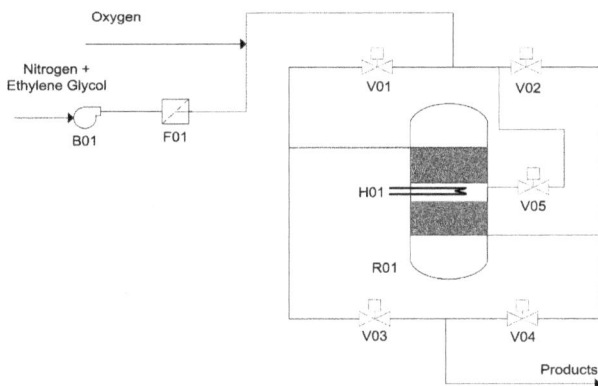

Figure 2. Sketch of the plant for the treatment of air-VOC mixtures in a reverse-flow reactor.

With respect to the alarms and control/protection systems, for the feeding/product lines, as well as for the gas blower and the filter, the same alarms previously described for the fixed-bed reactor based plant are used also in this case. The alarms and the control/protection systems used for the reactor are different, according to the system proposed by Barresi et al. [2] and based on the temperature measurement at the interface between the inert and the catalyst sections, where the reactor temperature is higher. In case this temperature is decreasing from the set-point value, when it reaches 410 °C, the heating capacity of H01 is set at the maximum value, the switching time is reduced and an alarm is activated. If the temperature continues decreasing, when it reaches the value of 390 °C a second alarm is activated and the shut-down of the reactor occurs. In case the temperature of the reactor increases, when it overtakes 500 °C a first protections system acts turning off the heater and increasing the switching time, and an alarm is activated. In case the temperature continues increasing, when it reaches the value of 530 °C another protection system is activated to protect the catalyst from the thermal degradation: in this case, the cold gas is feed directly in the middle of the reactor (opening the valve V05, and closing valves V01 and V02), and an alarm is activated.

3. Materials and Methods

3.1. Integrated Dynamic Decisional Analysis

3.1.1. Logical Model

The logical-probabilistic model is built starting from the process plant functional analysis, since it allows identifying and describing the events that could occur in the plant. Each event is included in the input model through a list of questions or affirmations, each of them characterized by the probability of occurrence for each level outcome, and, if available, by an uncertainty ratio, representing the distribution of the probability. Then a network is built indicating the subsequent level to be visited depending on the level outcome.

Figure 3 shows the structure of the logical-probabilistic analysis, declined for the sensing element AE01 described at level 10, characterized by a probability of failure of 0.062. The numbers that follow represent the levels to be visited in case of correct functioning and in case of failure of the component analyzed: in this case the level 11 in case of positive outcome and level 15 in case of failure.

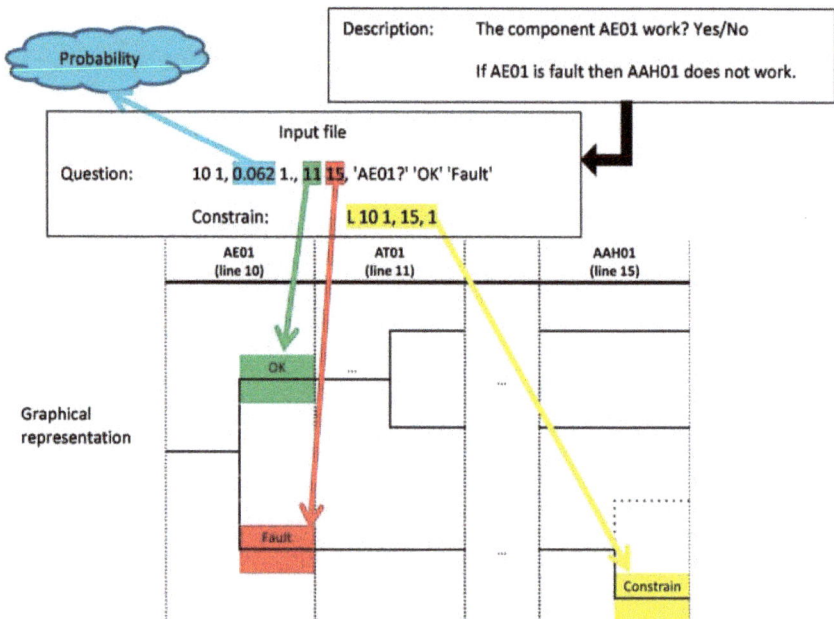

Figure 3. Graphical representation of the logical analyses structure.

When needed, each level is accompanied by a row defining a logical and probabilistic constraints, in order to take into account how an event can influence the subsequent events. In the example of Figure 3, in case of negative outcome of level 10 (10 1), the constraint forces the subsequent level to be visited (15) to the negative outcome also (1). This meaning that in case of failure of the sensing element, not only the control loop will fail, but also the alarm loop that depends on AE01.

The logical-probabilistic model is then built with IDDA 2.0 software (Software Oriented System Engineering S.r.l., Milan, Italy), elaborating all the possible sequences of events in the plant.

For both alternative process plants, the logical-probabilistic models were developed in the same way, by the systematic analysis of the equipment failure effects on the plant and of the protective devices and alarms effectiveness. The events occurrence probability was obtained by literature [29] and, where known, from the plant management. For an unbiased comparison of the two plants, the failure probability for similar pieces of equipment was considered to be the same.

For the fixed bed plant 3,901,910 sequences were generated and, with a cuff-off value of 10^{-16} a residual probability of 7.5×10^{-11} was obtained. The cut-off value is a probability threshold: the sequences with probabilities lower than the cut-off value are neglected and the sum of the probabilities of the neglected sequences gives the residual probability.

For the reverse-flow plant analysis a cut-off value of 10^{-12} was used and 5,336,624 sequences of events were obtained, with a residual probability of 2.9×10^{-6}.

The cut-off value is obtained through an optimization on the residual probability value, to be minimized against the number of sequences to be included in the calculation with respect to the available computational resources.

Among all the possible consequences, the more critical events from the point of view of the process (Top Events) were selected, namely the discharge with an excess of VOCs and the catalyst sintering. In Table 1 the occurrence probability for the Top Events is shown for the two plants analyzed. With respect to catalyst sintering, the logical-probabilistic analysis for the reverse flow plant did not identify any sequence resulting in a probability of occurrence higher than the cut-off value. The sequences with 3 or less equipment failures, or undesired outcomes, were extracted and elaborated jointly with the phenomenological model.

Table 1. Results of the logical–probabilistic model.

Top Event	Fixed Bed Plant		Reverse Flow Plant	
	Sintering of Catalyst	Discharge with Excess of VOCs	Sintering of Catalyst	Discharge with Excess of VOCs
Number of sequences	1,088,431	1,955,342	0	1,134,625
Probability	5.41×10^{-6}	8.37×10^{-3}	-	1.32×10^{-2}
Cut-off	10^{-16}	10^{-16}	10^{-12}	10^{-12}

3.1.2. Phenomenological Model

A mathematical model has been used to simulate plant dynamics. For the fixed-bed reactor based plant the main features are summarized in the following:

- Compressor (B01): the outlet pressure has been calculated evaluating the pressure drop in the plant; adiabatic compression has been assumed to calculate the outlet temperature and the compression work.
- Filter (F01): in case the filter operates correctly, the outlet flow rate is not modified, while in case the filter is clogged, only a fraction of the gas (assumed to be constant and equal to 50%) is assumed to pass through the filter, while the rest accumulates upstream the filter increasing the differential pressure.

- Oxygen input: aiming to maintain constant the oxygen excess in the reactor, the flow rate of oxygen is proportional to the difference between the oxygen concentration in the exhaust gas and the target value.
- Heat recovery device (H01): the heat balance is solved to calculate the outlet temperatures for the cold and hot sides. The global heat exchange coefficient (U, W/m^2·K) is calculated using the following equation:

$$\frac{1}{U} = \frac{1}{h_c} + \frac{1}{h_h} + \frac{s}{k}$$ (1)

where h (W/m^2·K) is the gas-wall heat transfer coefficient (h_h for the hot side and h_c for the cold side), k (W/m·K) is the wall thermal conductivity, and s (m) is the wall thickness. The gas-wall heat transfer coefficient is calculated using the following equation:

$$Nu = \frac{hD_e}{k_g} = 0.28Re^{0.65}Pr^{0.4}$$ (2)

where Nu is the Nusselt number, Re is the Reynolds number, and Pr is the Prandtl number, D_e (m) is the characteristic length (in our case, the equivalent diameter), k_g (W/m·K) is the gas thermal conductivity. The dirtying of the heat exchanger has been simulated considering a reduction of 30% of the heat exchange coefficient.

- Heater (H02): given the amount of heat provided by the heater, the outlet gas temperature is calculated from the heat balance. In case of failure of the control system of the heater, either the heater gives the highest heating rate (100 kW), and the gas temperature can be calculated accordingly, or the heater does not give any heat, and the exit temperature is equal to the inlet value.
- Reactor (R01): the set of equations used to calculate the dynamics of the reactor is composed by the energy balance for the gas:

$$\rho_g u_g c_{p,g} \varepsilon_c \frac{dT_g}{dz} - h_{gs}a_v(T_s - T_g) = 0$$ (3)

where ρ_g (kg/m^3) is the density of the gas, u_g (m/s) is the velocity of the gas, $c_{p,g}$ (J/kg·K) is the specific heat of the gas, ε_c (m^3/m^3) is the porosity of the bed, T_g (K) is the temperature of the gas, z (m) is the axial coordinate, h_{gs} (W/m^2·K) is the gas-solid heat transfer coefficient, a_v (m^2/m^3) is the specific surface of the catalyst, and T_s (K) is the solid temperature, by the energy balance for the solid:

$$\rho_s c_{p,s}(1 - \varepsilon_c)\frac{dT_s}{dt} = \frac{d^2T_s}{dz^2}k_s(1 - \varepsilon_c) - h_{gs}a_v(T_s - T_g) - (-\Delta H_r)\rho_g \varepsilon_c r_{VOC}$$ (4)

where ρ_s (kg/m^3) is the density of the solid, $c_{p,s}$ is the specific heat of the gas (J/kg·K), t (s) is the time, k_s (W/m·K) is the thermal conductivity of the solid, ΔH_r (J/kg) is the heat of reaction, and r_{VOC} (s^{-1}) is the specific reaction rate of the VOC) and by the mass balance for the VOC. In this case, the chemical reaction was considered as an instantaneous and complete reaction if the temperature is higher than the reaction temperature. Thus, in presence of an excess of oxygen the following equations are solved:

$$\begin{cases} y_{VOC} = 0 \\ \frac{dy_{O_2}}{dt} = \frac{dy_{O_2}}{dz}u_g - r_{VOC}\frac{2.5M_{O_2}}{M_{VOC}} \end{cases}$$ (5)

while in case of an excess of VOC the following equations are solved:

$$\begin{cases} y_{O_2} = 0 \\ \frac{dy_{VOC}}{dt} = \frac{dy_{VOC}}{dz}u_g - r_{VOC} \end{cases}$$ (6)

where y_{VOC} (kg/kg) is the VOC mass fraction, y_{O2} (kg/kg) is the oxygen mass fraction, M_{VOC} (kg/kmol) is the VOC molar mass, M_{O2} (kg/kmol) is the oxygen molar mass. In case the temperature of the solid is lower than the reaction temperature no chemical reaction occurs and, thus, the following equations are solved:

$$\begin{cases} \frac{dy_{O_2}}{dt} = \frac{dy_{O_2}}{dz} u_g \\ \frac{dy_{VOC}}{dt} = \frac{dy_{VOC}}{dt} u_g \end{cases} \tag{7}$$

Previous equations require adequate boundary conditions: the temperature and the composition of the gas entering the reactor are equal to the values obtained for the gas leaving the heater H01, and the solid does not exchange heat (by conduction) in correspondence of the two reactor edges.

For the reverse-flow reactor based plant the model of the compressor, of the filter, and of the oxygen input are nor modified. With respect to the reactor, for the catalytic part the equations solved are equal to Equations (1)–(7). For the inert sections the model is composed by the energy balance for the gas phase:

$$\rho_g u_g c_{p,g} \varepsilon_i \frac{dT_g}{dx} - h_i a_{v,i}(T_i - T_g) = 0 \tag{8}$$

where ρ_i (kg/m^3) is the inert density, $c_{p,i}$ (J/kg·K) is the inert specific heat, u_g (m/s) is the velocity of the gas, ε_i (m^3/m^3) is the void fraction of the bed of inert, h_i (W/m^2·K) is the gas-inert heat transfer coefficient, $a_{v,i}$ (m^2/m^3) is the specific surface area of the inert, and T_i (K) is the temperature of the inert, T_g (K) is the temperature of the gas, k_i (W/m·K) is the thermal conductivity of the inert, and by the heat balance for the solid:

$$\rho_i c_{p,i}(1 - \varepsilon_i)\frac{dT_i}{dt} = \frac{d^2 T_i}{dx^2} k_i(1 - \varepsilon_i) - h_i a_{v,i}(T_i - T_g) \tag{9}$$

and by Equation (7) as no chemical reaction takes place in the inert sections.

The electric heater in the central section, used for control purposes, is modeled using a heat balance for the gas phase:

$$Q_h = \rho_g u_h c_{p,g} S(T_{g,2} - T_{g,1}) \tag{10}$$

where Q_h (W) is the heat supplied by the heater, $T_{g,2}$ (K) is the gas temperature after the heater, $T_{g,1}$ (K) is the gas temperature before the heater e S (m^2) is the section of the reactor.

The phenomenological model developed in Matlab according to the above described equations gives results coherent with the literature results.

Consequence Assessment and CBA

The operational costs and the time losses in case of plant malfunction were used to compare the two plants. The operational costs take into account the recovery of the failed pieces of equipment, the costs related to the plant stops, the cost related to the VOC discharge and the energy consumption. The time lost takes into account the production losses due to plant shut down, for the equipment restoration and the plant start up. Table 2 shows for each major intervention in case of fault or any major consequence, both the costs of the operation and the production time lost, as provided by the plant managers.

Each sequence of events is coupled with a consequence value and the model of its phenomenological behavior. Table 3 shows a sample sequence, where the lack of oxygen in the reactor is due to the lack of oxygen main source, thus bringing to a low temperature in the reactor itself; following the low temperature the sequence includes the failure of the lower temperature protection system (TSL10), but the emergency state is managed thanks to an alarm (TAL10) and the operator intervention. The Table shows also the probability of occurrence of this sequence of events and its consequence. In the case presented in Table 3, the operational cost was evaluate as the sum of the oxygen input system restoration (2000 €), the cost of the emergency shut down (100,000 €) and the cost of the energy used in the plant for the year taken into account (26,500 €). The risk value for an unwanted event is the sum of the risks of the sequences of events bringing to the unwanted event itself. The risk value is thus evaluated with Equation (11):

$$R = \sum C_i \cdot P_i \tag{11}$$

where C_i is the consequence of the i-th sequence of events and P_i is its probability.

Table 2. Summary of cost and time loss.

Intervention	Estimated Costs	Estimated Time
Restore filter	200 €	3 h
Restore system of input of oxygen	2000 €	2 h
Restore heat recovery	20,000 €	5 days
Restore heater	10,000 €	1 day
Restore blower	2000 €	1 day
Restore valve of reactor	10,000 €	2 days
Restore heat control inside of reactor	2000 €	3 days
Replace catalyst in traditional plant	200 €/dm^3	7 days
Replace catalyst in intensified plant	17,500 €	7 days
Plant stop	100,000 €	1 day
Cost no abatement of VOCs	60,000 €	-
Cost of electric power	0.16 €/kWh	-

Table 3. Example of an events' sequence.

Sequences Number 10119			Probability 2.43 × 10^{-9}		
Out Service	26 h/Year		Risk (out of Service)		6.31 × 10^{-8} h/Year
Operational Cost	128,500 €/Year		Risk (Operational Cost)		0.0003 €/Year
Level	Probability	Cumulative Probability	Description	Out Service (h/Year)	Cost (€/Year)
1	1-1.00 × 10^{-3}	0.999	Input temperature? OK	-	-
3	1-1.00 × 10^{-3}	0.998	Input VOCs concentration? OK	-	-
5	1-1.00 × 10^{-3}	0.997	Input Flow? OK	-	-
50	1-8.40 × 10^{-2}	0.913	Blower B01? OK	-	-
52	1-9.50 × 10^{-2}	0.827	FE01? OK	-	-
53	1-2.52 × 10^{-1}	0.618	FIC01? OK	-	-
54	1-2.31 × 10^{-1}	0.475	FCV01? OK	-	-
55	-	0.475	Flow after the blower? OK	-	-
100	1-9.00 × 10^{-1}	0.0475	Filter F01? OK	-	-
150	1-6.20 × 10^{-2}	0.0446	AE02 (O2)? OK	-	-
151	1-9.50 × 10^{-2}	0.040357	AT02? OK	-	-
152	1-2.52 × 10^{-1}	0.030187	AICA02? OK	-	-
153	1-1.81 × 10^{-1}	0.024723	FE02? OK	-	-
154	1-9.50 × 10^{-2}	0.022375	FT02? OK	-	-
155	1-2.52 × 10^{-1}	0.016736	FIC02? OK	-	-
156	1-2.31 × 10^{-1}	0.01287	FV02? OK	-	-
157	0.001	1.29 × 10^{-5}	O2 is Available? NO	2	2.000
160	-	1.29 × 10^{-5}	The O2 concentration? Low	-	-
230	1-7.00 × 10^{-3}	1.28 × 10^{-5}	V01? OK	-	-
232	1-7.00 × 10^{-3}	1.27 × 10^{-5}	V02? OK	-	-
234	1-7.00 × 10^{-3}	1.26 × 10^{-5}	V03? OK	-	-
236	1-7.00 × 10^{-3}	1.25 × 10^{-5}	V04? OK	-	-
245	-	1.25 × 10^{-5}	Flow from V01 to V04? Correct	-	-
255	-	1.25 × 10^{-5}	Flow from V02 to V03? Correct	-	-
276	-	1.25 × 10^{-5}	Temperature inside reactor? Low	-	-
500	1-8.40 × 10^{-2}	1.15 × 10^{-5}	TE10? OK	-	-
502	1-2.52 × 10^{-1}	8.57 × 10^{-6}	TIC10? OK	-	-
521	0.0003	2.57 × 10^{-9}	TSL10? Fault	-	-
528	-	2.57 × 10^{-9}	The system TSL10 is effectiveness? NO	-	-
504	1-2.50 × 10^{2}	2.51 × 10^{-9}	TAL10? OK	-	-
505	1-3.00 × 10^{-2}	2.43 × 10^{-9}	Operator occur on TAL10? YES	-	-
506	-	2.43 × 10^{-9}	TLA10 is effectiveness? YES-Emergency	-	-
602	-	2.43 × 10^{-9}	Plant status? Emergency	24	100.000
-	-	-	Power		26.500

4. Results and Discussion

In this section, the results of the logical-probabilistic and phenomenological modelling are discussed.

Figure 4 shows the probability of occurrence of the sequences bringing to the discharge with an excess of VOCs, also represented in Figure 5 in terms of risk; Figure 6 shows the 100 more relevant sequences of events ((a) for the out of service risk and (b) for the operational cost risk).

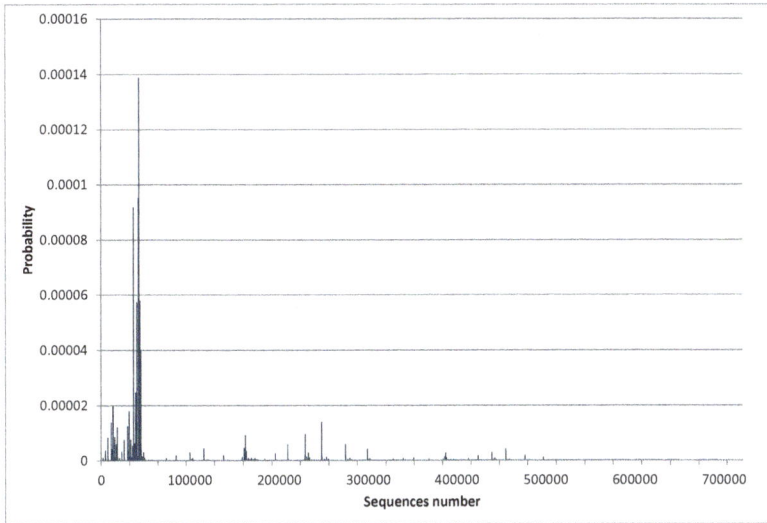

Figure 4. The probability of the sequences of events contain the high concentration of VOCs in the output for the fixed-bed based plant.

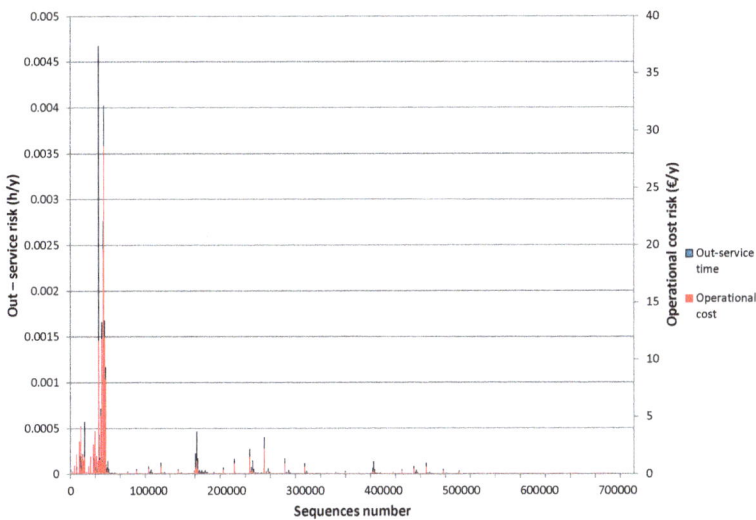

Figure 5. The risk of the sequences of events contain the high concentration of VOCs in the output for the fixed-bed based plant.

Both Figures highlight how a small number of sequences mainly contributes to the global probability or risk. Those major contributors are the sequences where the oxygen input fails or the heater fails, without the intervention of the alarm and/or of the protective systems. It can be observed that the operational cost risk has the same trend of the probability, while for the out-of-service risk the relevance of the sequences is different with respect to the probabilities. In fact, the main contributors are those pertaining the failure of the heater, while those involving the oxygen feed failure are less relevant.

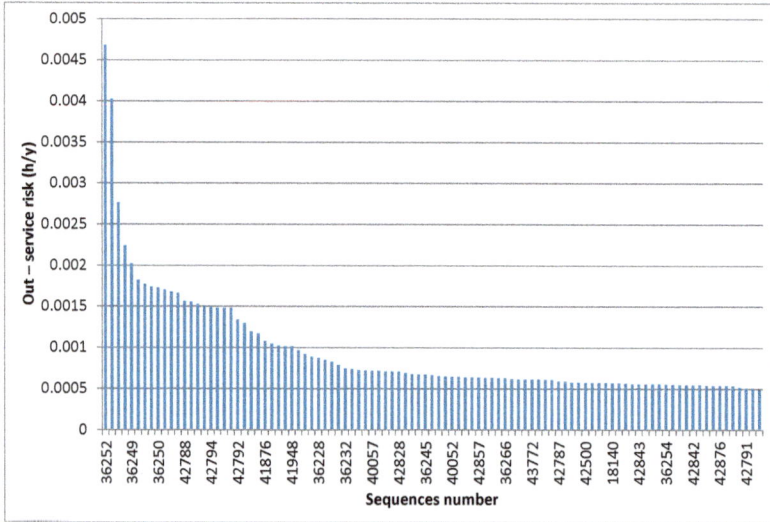

(a)

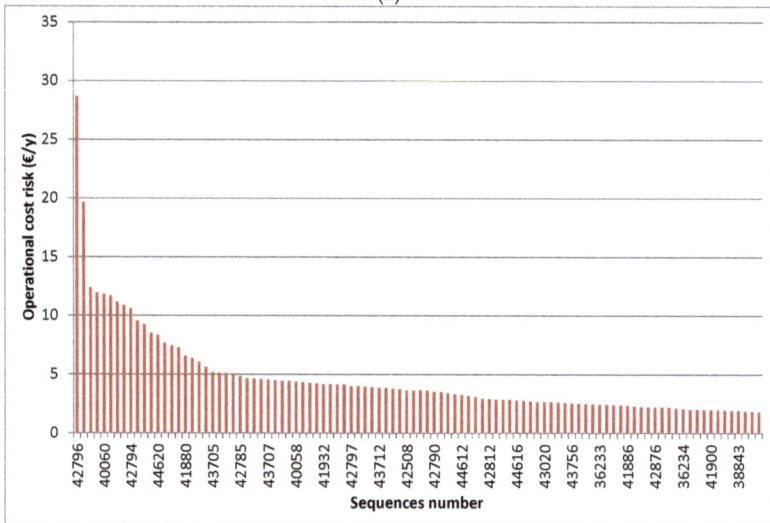

(b)

Figure 6. Details of the results bringing to high concentrations of VOCs in the output stream for the fixed bed based plant (**a**) The 100 sequences with the higher value of out-service time risk; (**b**) The 100 sequences with the and higher value of risk of operational cost.

Figure 7 shows the distribution of the probabilities bringing to the discharge with an excess of VOCs, while Figure 8 represents the risk associated to the same sequences risk; Figure 9 shows the 100 more relevant sequences of events ((a) for the out-service risk and (b) the operational cost risk). Also for the reverse-flow based plant a small number of sequences brings the higher contribution to the global value of probability or risk. These sequences include the fault of the oxygen input system and differ for the system's component failing. The probability and the risk values show the same trend.

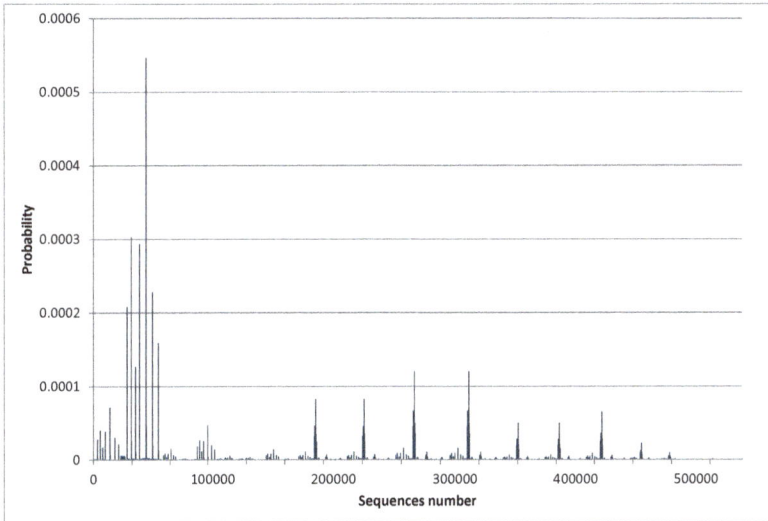

Figure 7. The probability of the sequences of events contain the high concentration of VOCs in the output for the reverse-flow based plant.

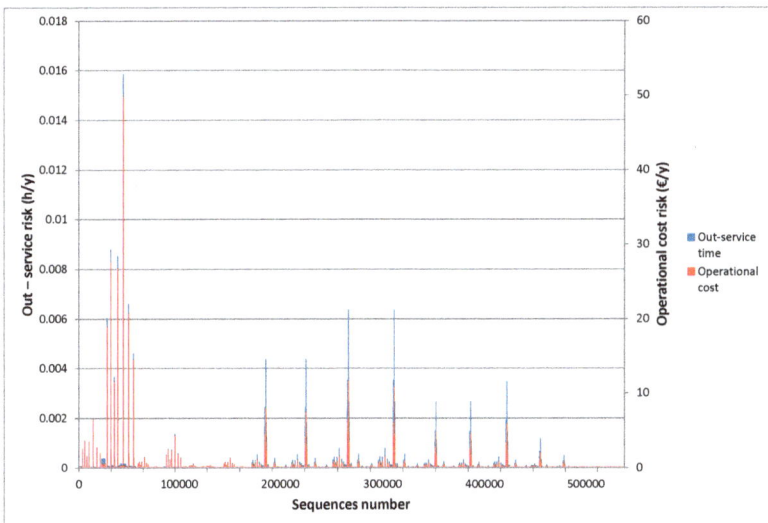

Figure 8. The risk of the sequences of events contain the high concentration of VOCs in the output for the reverse flow plant.

(a)

(b)

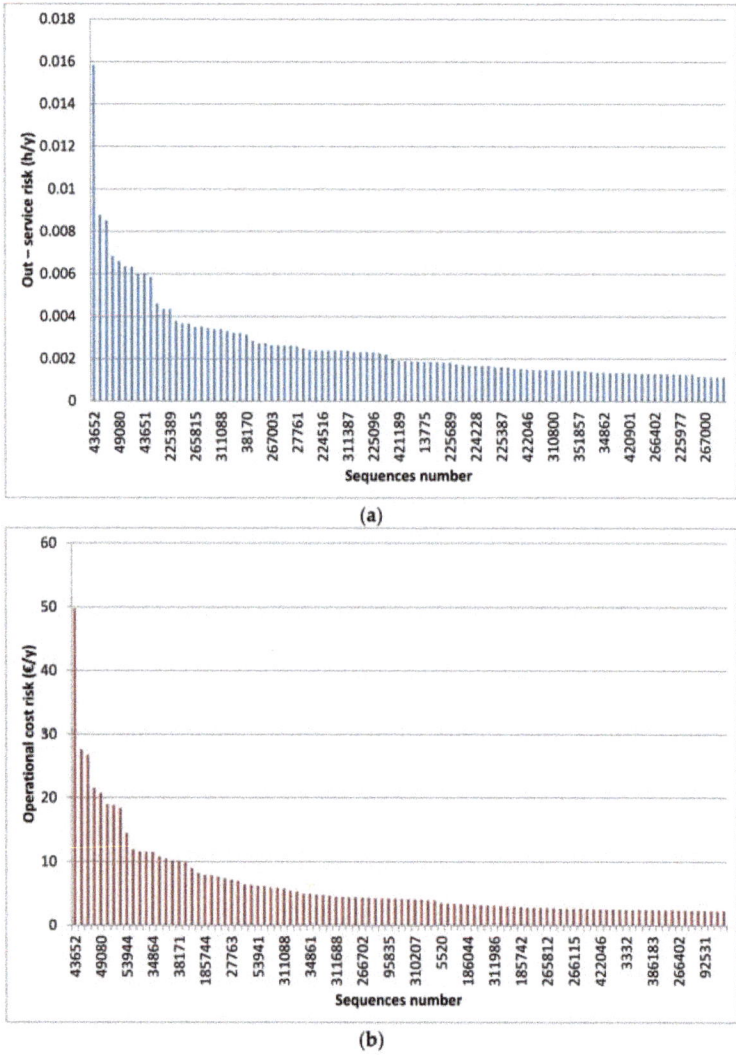

Figure 9. Details of the results bringing to high concentrations of VOCs in the output stream for the revers flow plant (**a**) The 100 sequences with the higher value of out-service time risk; (**b**) The 100 sequences with the and higher value of risk of operational cost.

From the comparison between the risk figures it is possible to conclude that the reverse-flow based plant is less risky than the fixed-bed based plant from the operational costs point of view, mainly thanks to the lower energy consumption. In terms of service outage, the reverse-flow based plant appears to be riskier than the fixed-bed based plant, because of its higher complexity, with a higher time needed for restoration after a failure.

Analyzing the sintering of the catalyst, that represents a critical event for the plant operations, the fixed-bed based plant shows a 5.41×10^{-6} probability of occurrence, while the reverse-flow based plant does not have sequences of events bringing to the sintering of the catalyst with a probability of occurrence higher than the cut off value (10^{-12}). In a precautionary way, all the sequences constituting the residual probability were assimilated to the catalyst sintering probability in the reverse-flow based plant, although this is an overestimation of the probability itself. In Figure 10 the probability distribution of the sequences of events for the catalyst sinterization in the fixed-bed based plant is shown. In Figure 11 the risk distribution is shown. In these two graphs the more relevant sequences include the filter obstruction combined to the fault of the heater control system. Figure 12 shows the 100 more relevant sequences of events ((a) for the out-service risk and (b) for the operational cost).

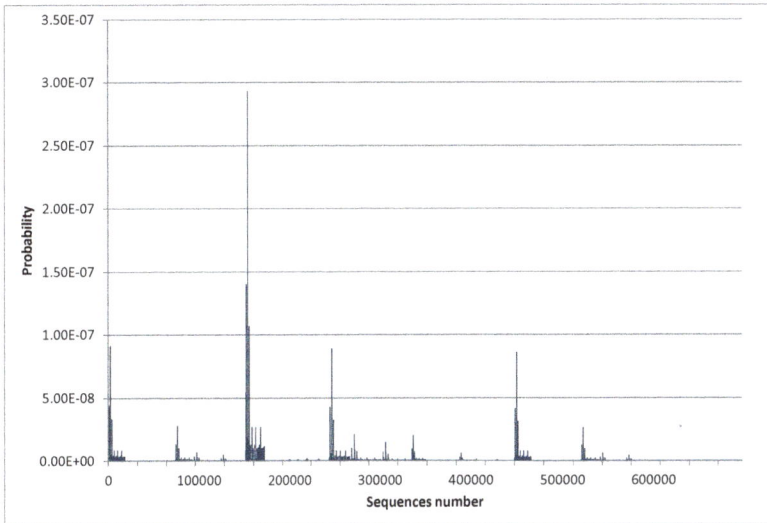

Figure 10. The probability of the sequences of events contain the catalyst sinterization for the fixed-bed based plant.

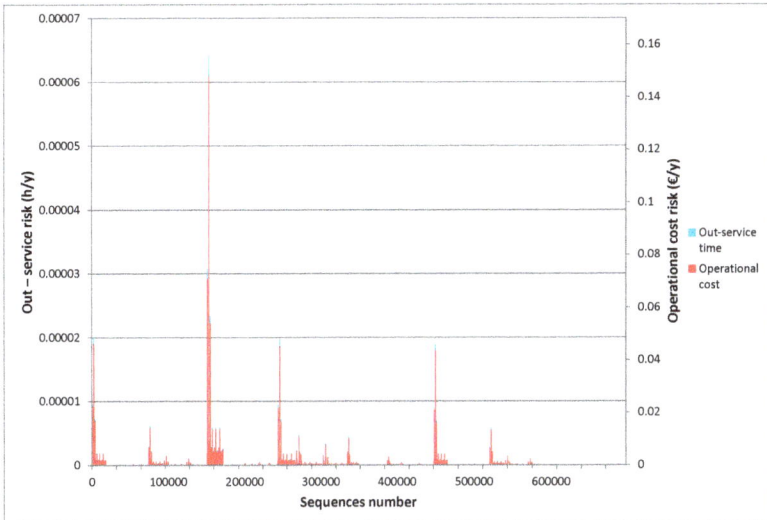

Figure 11. The risk of the sequences of events contain the catalyst sinterization for the fixed-bed based plant.

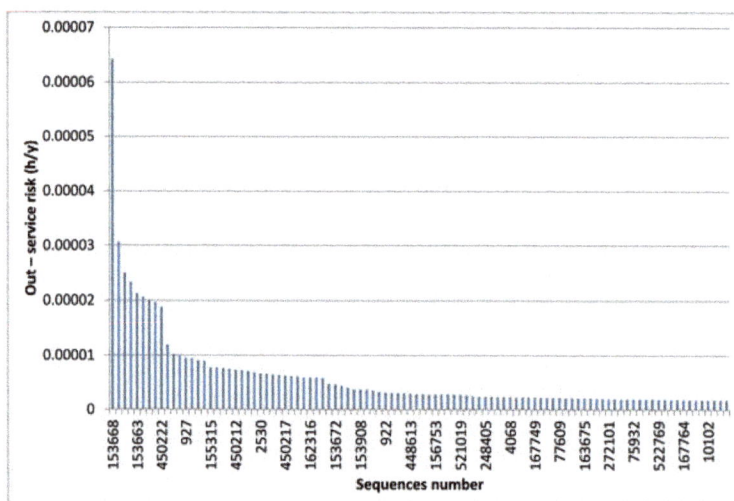

(a)

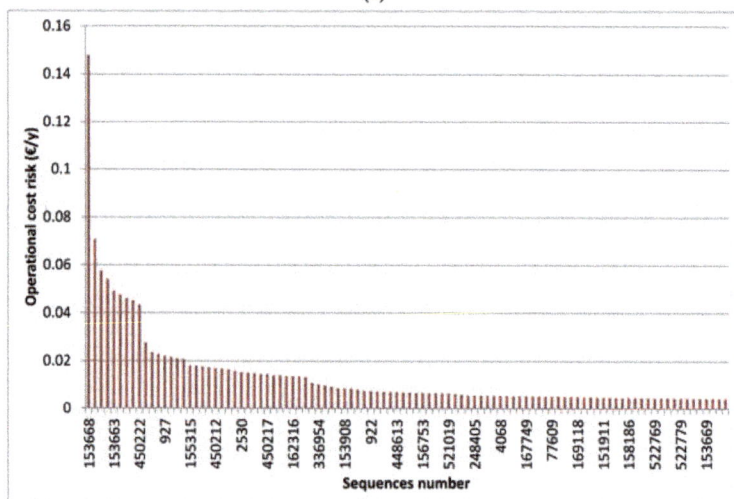

(b)

Figure 12. Details of the results bringing to the catalyst sinterization for the fixed bed based plant (**a**) The 100 sequences with the higher value of out-service time risk; (**b**) The 100 sequences with the and higher value of risk of operational cost.

Table 4 shows the summary of the results used for the risk based decision-making. It also shows the mean values of the annual operation cost and the time losses evaluate for the sequences taken into account. The discharge with excess of VOCs causes an unwanted shut down of the plant, with possible process problems on the subsequent systems. The fixed bed plant appears to be more reliable than the reverse flow plant (8.37×10^{-3} vs. 1.32×10^{-2} of probability).

The risk of discharge with excess of VOCs for the fixed bed plant corresponds to 1400 €/year and 0.35 h/year in terms of the time losses. For the reverse-flow based plant, instead, the risk figures are 1200 €/year and 0.56 h/year.

Table 4. Fixed bed plant and reverse flow plant results.

Indicator	Fixed Bed Plant		Reverse Flow Plant	
	Sintering of Catalyst	Discharge with Excess of VOCs	Sintering of Catalyst	Discharge with Excess of VOCs
Probability	5.41×10^{-6}	8.37×10^{-3}	2.9×10^{-6} [1]	1.32×10^{-2}
Annual operational costs, €/year	500,000	171,000	144,000	92,000
Risk, €/year	2.7	1400	0.4	1200
Time losses, h/year	219	42	192	43
Risk, h/year	1.2×10^{-3}	0.35	5.6×10^{-4}	0.56
Number of sequences	698,688	721,461	0	538,590

[1] This probability value is the residual probability value used in conservative way for probability of the event here described for the plant intensified.

5. Conclusions

In the case study for the selection of competing technologies for the VOCs treatment, from the analysis of the results it is possible to conclude that, depending on the benefits the stakeholder is interested in, the following conclusions can be drawn.

Considering the time losses following a failure, the two plant alternatives appear to be comparable in terms both of consequences and of risk of discharge with an execs of VOCs. For both plants the more critical system appear to be the input of oxygen, on which the inspections and maintenance activities should be concentrated.

When the benefits related to the minimization of the operational costs are taken into account, instead, the intensified plant (reverse-flow reactor) is the best option, since it can run with a less expensive catalyst and it requires less energy.

Thus, given the above considerations, the reverse flow reactors appear to be the most effective technology to treat lean-VOC streams.

The different risk figures allow the management to have a global view of the plant foreseeable behavior and to take better and more informed decisions.

The subsequent decision depends on the final aim of the plant managers, that have in any case available the full picture of the behavior of the plant alternatives.

The Integrated Dynamic Decisional Analysis appears to be an effective way to help the management to perform a risk based decision between different technological solution The IDDA analysis allow analyzing in a systematic way the plant behavior in case of normal condition and in case of process deviation and/or equipment failure. It also allows to evaluate the occurrence probability of the unwanted outcomes, but also their risk value in a cost-benefit shape.

Author Contributions: The three authors jointly conceived the paper based on the case study of industrial interest. Davide Fissore developed the phenomenological model, implemented by Gabriele Baldissone. Micaela Demichela and Gabriele Baldissone developed and implemented the logical and probabilistic model. All the authors jointly analysed the results and wrote the paper.

Conflicts of Interest: The authors declare no conflict of interest.

References

1. Kolaczkowski, S. Treatment of Volatile Organic Carbon. Emissions from stationary sources: Catalytic oxidation of the gaseous phase. In *Structured Catalysts and Reactors*, 2nd ed.; CRC Press: Boca Raton, FL, USA, 2006.
2. Barresi, A.A.; Baldi, G.; Fissore, D. Forced unsteady-state reactors as efficient devices for integrated processes: Case histories and new perspectives. *Ind. Eng. Chem. Res.* **2007**, *46*, 8693–8700. [CrossRef]
3. Cottrell, F.G. Purifying Gases and Apparatus Therefore. U.S. Patent 2,171,733, 21 June 1938.
4. Eigenberger, G.; Nieken, U. Catalytic combustion with periodical flow reversal. *Chem. Eng. Sci.* **1988**, *43*, 2109–21198. [CrossRef]
5. Nieken, U.; Eigenberger, G. Fixed-bed reactors with periodic flow reversal: Experimental results for catalytic combustion. *Catal. Today* **1994**, *38*, 335–350. [CrossRef]

6. Matros, Y.S.; Bunimovich, G.A. Reverse-flow operation in catalytic reactors. *Catal. Rev.* **1996**, *38*, 1–68. [CrossRef]

7. Van de Beld, L.; Westerterp, K.R. Air purification in a reverse-flow reactor: Model simulations vs. experiments. *AIChE J.* **1996**, *42*, 1139–1148. [CrossRef]

8. Zufle, H.; Turek, T. Catalytic combustion in a reactor with periodic flow reversal: 1. Experimental results. *Chem. Eng. Process. Process Intensif.* **1997**, *36*, 327–339. [CrossRef]

9. Fissore, D.; Barresi, A.A.; Baldi, G.; Hevia, M.A.G.; Ordònez, S.; Dìez, F.V. Design and Testing of Small-Scale Unsteady-State Afterburners and Reactors. *AIChE J.* **2005**, *51*, 1654–1664. [CrossRef]

10. Cittadini, M.; Vanni, M.; Barresi, A.A.; Baldi, G. Reverse-Flow catalytic burners: Response to periodical variations in the feed. *Chem. Eng. Sci.* **2001**, *56*, 1443–1449. [CrossRef]

11. Chen, G.; Chi, Y.; Yan, J.; Ni, M. Effect of periodic variation of the inlet concentration on the performance of reverse flow reactors. *Ind. Eng. Chem. Res.* **2011**, *50*, 5448–5458. [CrossRef]

12. Marín, P.; Ordóñez, S.; Díez, F.V. Monoliths as suitable catalysts for reverse-flow combustors: Modeling and experimental validation. *AIChE J.* **2011**, *56*, 3162–3173. [CrossRef]

13. Kolios, G.; Frauhammer, J.; Eigenberger, G. Autothermal fixed-bed reactor concepts. *Chem. Eng. Sci.* **2000**, *55*, 5945–5967. [CrossRef]

14. Luzi, C.D.; Martínez, O.M.; Barreto, G.F. Autothermal reverse-flow reactors: Design and comparison of valve-operated and rotary systems. *Chem. Eng. Sci.* **2016**, *148*, 170–181. [CrossRef]

15. Demichela, M.; Piccinini, N. Integrated dynamic decision analysis: A method for PSA in dynamic process system. In Proceedings of the CISAP 3, Rome, Italy, 11–14 May 2008; AIDIC: Milano, Italy, 2008.

16. Demichela, M.; Camuncoli, G. Risk based decision making. Discussion on two methodological milestones. *J. Loss Prev. Process Ind.* **2014**, *28*, 101–108. [CrossRef]

17. Leva, M.C.; Pirani, R.; Demichela, M.; Clancy, P. Human factors issues and the risk of high voltage equipment: Are standards sufficient to ensure safety by design? *Chem. Eng. Trans.* **2012**, *26*, 273–278.

18. Demichela, M.; Pirani, R.; Leva, M.C. Human factor analysis embedded in risk assessment of industrial machines: Effects on the safety integrity level. *Int. J. Perform. Eng.* **2014**, *10*, 487–496.

19. Saarikoski, H.; Mustajoki, J.; Barton, D.N.; Geneletti, D.; Langemeyer, J.; Gomez-Baggethun, E.; Marttunen, M.; Antunes, P.; Keune, H.; Santos, R. Multi-Criteria Decision Analysis and Cost-Benefit Analysis: Comparing alternative frameworks for integrated valuation of ecosystem services. *Ecosyst. Serv.* **2016**, *22*, 238–249. [CrossRef]

20. Dixon, J.A.; Hufschmidt, M. *Economic Valuation Techniques for the Environment. A Case Study Workbook*; Johns Hopkins University Press: Baltimore, MA, USA, 1986.

21. Boardman, A.; Greenberg, D.; Vining, A.; Weimer, D. *Cost-Benefit Analysis. Concepts and Practice*; Pearson Prentice Hall: Upper Saddle River, NJ, USA, 2011.

22. Ferreira, S.; Cabral, M.; da Cruz, N.F.; Simões, P.; Marques, R.C. The costs and benefits of packaging waste management systems in Europe: The perspective of local authorities. *J. Environ. Plan. Manag.* **2017**, *60*, 773–791. [CrossRef]

23. Nichol, KL. Cost-Benefit Analysis of a Strategy to Vaccinate Healthy Working Adults against Influenza. *Arch. Intern. Med.* **2001**, *161*, 749–759. [CrossRef] [PubMed]

24. Diakoulaki, D.; Karangelis, F. Multi-criteria decision analysis and cost–benefit analysis of alternative scenarios for the power generation sector in Greece. *Renew. Sustain. Energy Rev.* **2007**, *11*, 716–727. [CrossRef]

25. Clementel, S.; Galvagni, R. The use of the event tree in the design of nuclear power plants. *Environ. Int.* **1984**, *10*, 377–382. [CrossRef]

26. Galvagni, R.; Clementel, S. Risk analysis as an instrument of design. In *Safety Design Criteria for Industrial Plants*; Cumo, M., Naviglio, A., Eds.; CRC Press: Boca Raton, FL, USA, 1989; Volume 1.

27. Turja, A.; Demichela, M. Risk based design of allyl chloride production plant. *Chem. Eng. Trans.* **2011**, *24*, 1087–1092.

28. Gerbec, M.; Baldissone, G.; Demichela, M. Design of procedures for rare, new or complex processes: Part 2—Comparative risk assessment and CEA of the case study. *Saf. Sci.* in press. [CrossRef]

29. Mannan, S. *Lee's Loss Prevention in the Process Industries*; Elsevier: Oxford, UK, 2005.

environments

MDPI

Article

Biotrickling Filtration of Air Contaminated with 1-Butanol

Thomas Schmidt and William A. Anderson *

Department of Chemical Engineering, University of Waterloo, Waterloo, ON N2L 3G1, Canada;
t4schmid@uwaterloo.ca
* Correspondence: wanderson@uwaterloo.ca

Received: 27 June 2017; Accepted: 10 August 2017; Published: 15 August 2017

Abstract: The removal of high concentrations of 1-butanol in an air stream was evaluated with a biotrickling filter for potential application to an industrial off-gas. Experiments were conducted on a laboratory-scale system, packed with perlite, in a co-current downward mode with constant recycling of water. The performance was monitored for different inlet concentrations and empty bed residence times during a period of over 60 days of stable operation. A maximum elimination capacity (EC) of $100 \text{ g m}^{-3} \text{ h}^{-1}$ was achieved during periods in which the butanol concentration varied from 0.55 to 4.65 g m^{-3}. The removal efficiency was stable and exceeded 80% for butanol concentrations in the range of 0.4 to 1.2 g m^{-3}, corresponding to inlet mass loadings of up to approximately $100 \text{ g m}^{-3} \text{ h}^{-1}$. However, when the concentration exceeded 4 g m^{-3}, removal efficiency rapidly dropped to 15% (EC of $22 \text{ g m}^{-3} \text{ h}^{-1}$), indicating an inhibition effect that was reversed by decreasing the inlet concentration. This biotrickling filter was able to deal with higher sustained butanol concentrations than have been previously reported, but might not be suitable for concentrations much in excess of 1.2 g m^{-3} or mass inlet loads in excess of $100 \text{ g m}^{-3} \text{ h}^{-1}$.

Keywords: biotrickling filter; biofiltration; inhibition; solvents; n-butyl alcohol; elimination capacity

1. Introduction

The emissions of volatile organic compounds (VOCs) into the environment need to be controlled appropriately to protect or improve local air quality and to minimize tropospheric ozone formation. Although several different technologies are available, biological air treatment has attracted interest, especially for applications with moderate VOC concentrations below 10 g m^{-3} [1,2], because of its potentially favourable economics [3]. In comparison to conventional pollution control technologies, such as absorption, catalytic oxidation, condensation, and incineration, biological air treatment appears to have advantages which include high removal efficiencies, low installation and operating costs, good reliability, stable performance, and applicability to situations with larger volumes of waste gases containing lower concentrations of VOCs. Numerous examples of successful applications in the treatment of VOCs and odours can be found in literature [4–6]. Much of the work has focused on biofiltration, where a humidified gas is routed through a microbe-containing packed bed. Other work has been reported on biotrickling filtration, where a liquid phase continuously flows over an inert packing material that supports the microbial community. Biotrickling filtration has several potential advantages for systems that are difficult to pre-humidify, or that present challenges in pH and/or temperature control [7,8].

In this work, control of 1-butanol (n-butyl alcohol) was of interest because it was identified as a major component of a local industrial intermittent emission resulting from thermal regeneration of adsorbents, at concentrations in the range of 5 to 10 g m^{-3} or even higher. Studies have been reported in literature with laboratory-scale bioreactors to evaluate the performance for a range of hydrocarbons and oxygenated hydrocarbons [9–16]. However, the biological treatment of 1-butanol in air has not been as extensively reported as some other compounds, especially at higher concentrations. Heinze and Friedrich [17] examined 1-butanol biofiltration (not biotrickling) with several different packing materials, but at relatively low concentrations (0.24 g m^{-3}). Fitch et al. [18] successfully treated air phase 1-butanol in a membrane bioreactor at concentrations up to approximately 1.1 g m^{-3}. Lee et al. [19] treated approximately 4.3×10^{-4} g m^{-3} 1-butanol in a gas phase mixture including acetone and ammonia. Ondarts et al. [20] evaluated a compost biofilter operation for a VOC mixture where butanol was present at concentrations up to 1.9×10^{-5} g m^{-3}. Chan and Lai [11] studied the interaction between 1-butanol and 2-butanol degradation in polyvinyl alcohol bead-packed biofilters at concentrations up to approximately 1 g m^{-3}, noting that 1-butanol inhibited 2-butanol degradation. Feizi et al. [21] tested higher concentrations of n-butanol concentrations (up to 3.2 g m^{-3}), but only under transient conditions lasting a few hours. One of the only biotrickling filter studies was performed by Wang et al. [22] on a mixture of butyl acetate, butanol and phenyl acetic acid at butanol concentrations up to 2.4 g m^{-3}.

Most of these prior studies were performed at concentrations well below those expected in this potential industrial application, and it is known that butanol is inhibitory or toxic to microbes at concentrations above 2% in liquid phase fermentations [23]. Therefore, the objective of this study was to investigate the removal performance for 1-butanol at higher sustained concentrations than have been reported previously, to determine if a biological treatment technology might be suitable for application under these conditions.

Humidification of the contaminated air was not expected to be feasible in this proposed industrial application, and so a biotrickling filter (BTF) operating mode was chosen whereby an aqueous solution was continuously trickled over the top of the bed where the butanol-contaminated air entered. In this manner, the inlet of the device could be kept moist while also serving to humidify the gases as they travelled down the bed. Biotrickling filters also provide a mechanism for potentially better temperature, pH, and nutrient concentration control than conventional biofilters. Perlite was chosen as the packing for this study due to its good mechanical and non-compactible properties, which were considered important for this specific application. Perlite is a readily available naturally occurring siliceous volcanic rock [24], with a porous surface that has yielded high volumetric elimination capacities in other biofilters and biotrickling filters [1,14,25].

2. Materials and Methods

Nutrient-enriched perlite was purchased from a local supplier (Miracle-Gro®Perlite 0.04-0.01-0.06 enriched, Scotts Canada Ltd., Mississauga, CA, USA). Material characteristics were reported to be: 0.04% total N, 0.01% available phosphorus (P$_2$O$_5$), 0.06% soluble potash (K$_2$O), 200 kg m^{-3} bulk density, and near neutral pH. The perlite was screened into two size fractions. In the upper (inlet) 1/8 section of the BTF, only perlite of a diameter greater than 4.7 mm was used to help prevent plugging of the column in this region where the most biomass growth was expected. The remaining 7/8 section contained perlite with a size between 4.7 and 1.7 mm.

The experimental work was performed using a laboratory-scale biotrickling filter (BTF) as shown in Figure 1, constructed from glass tubing with an internal diameter of 11.5 cm and a total height of 90 cm. The BTF was equipped with sampling ports to allow sampling of the stream entering and leaving the BTF, as well as five ports (1–5) axially along the medium bed for gas samples and 3 ports (A–C) for packing material samples. The BTF was packed with perlite to a depth of 67 cm, resulting in a working bed volume of 6.9 L.

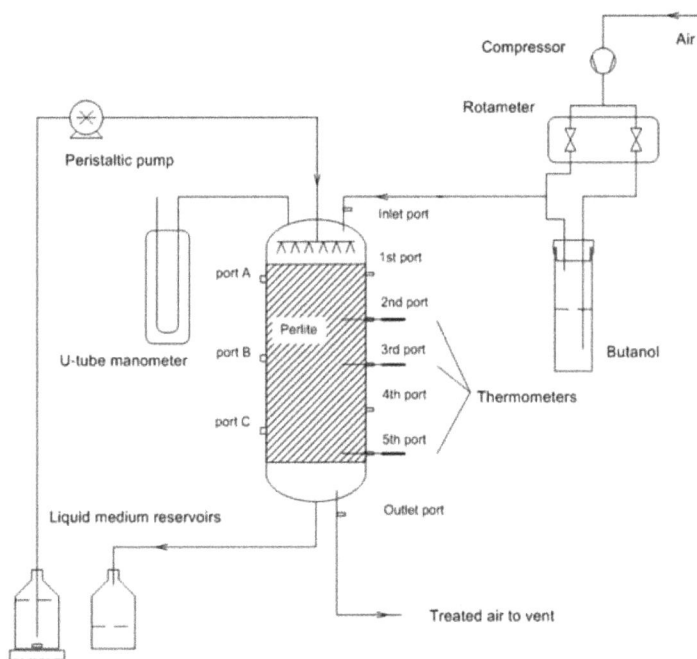

Figure 1. Schematic diagram of the experimental system.

For start-up, a mixed microbial inoculum was prepared and acclimated by placing a sample of a local forest soil in tap water and aerating the solution for 3 days with an air stream containing butanol. Sodium acetate (5 g L^{-1}) was added to provide an initial carbon source that was more readily utilized, and a few drops of soybean oil minimized the formation of foam. At start up, the BTF bed was inoculated over a period of 24 h by recirculation of 2.5 L of the acclimated solution at a flow rate of 40 mL min^{-1} to distribute microorganisms over all of the packing material. The BTF was operated in a co-current gas and liquid downward flow mode to simplify hydration of the packing in the entrance region where drying tends to occur most rapidly. A continuous flow of water was trickled onto the top of the filter bed through nozzles. The drainage solution was collected at the bottom of the column in a 4 L vessel and recycled using a peristaltic pump at a flow rate of 40 mL min^{-1}. The water was supplemented on several days (27 and 40) with a mineral salt medium consisting of (g L^{-1}): Na$_2$HPO$_4$ 3, (NH$_4$)$_2$SO$_4$ 1.5, NH$_4$Cl 3, KNO$_3$ 4.5, CaCO$_3$ 1.5, MgSO$_4$ 1.5, K$_2$SO$_4$ 1.5, in deionised water [6,26]. The pH of the solution was neutral and the calcium carbonate served as a buffering material.

The air stream was divided into two, with the smaller stream (approximately 10 to 20% of the total) directed to a vial where it bubbled through about 10 cm of liquid butanol maintained at an approximately constant volume. The butanol vapourized into the air stream, and was mixed with the second air stream and introduced into the column. The concentration of butanol in the air stream could be adjusted to the desired value by changing the ratio of the air streams, whereby the total volumetric flow stayed constant. The pressure drop across the packing material was measured by a differential U-tube water manometer connected to the top of the column. The temperature and relative humidity of the air stream was monitored with a thermohygrometer (Oakton, Vernon Hills, IL, USA). The relative humidity of the inlet air ranged between 8 and 10%, while the temperature did not deviate significantly from 23 °C. During the experiments, the inlet gas flowrate (and corresponding empty bed residence time, EBRT) was set at 0.20 m^3 h^{-1} (days 1 to 6, EBRT 124 s), 0.277 m^3 h^{-1} (days 7 to 41, EBRT 90 s), 0.415 m^3 h^{-1} (days 42 to 50, EBRT 60 s), and 0.277 m^3 h^{-1} (final 5 days, EBRT 90 s).

Gas samples for VOC and CO_2 analysis were taken with a 1 mL gas-tight syringe. The concentration of butanol was determined by gas chromatography (GC) (HP 5890 Series II, Agilent, Santa Clara, CA, USA) with a flame ionization detector, using a RTX-502.3 fused silica megabore column of dimensions 30 m × 0.53 mm and 3 μm film thickness. The initial oven temperature was 50 °C for 3 min, increasing at 20 °C min^{-1} to 80 °C where it was held for 1.5 min. Gas phase samples for CO_2 analysis were also taken through the sampling ports at the top and bottom of the BTF. A second GC (HP 5890 Series II) was used for determining the CO_2 concentration using He carrier and make-up gas, with a Porapak Q column (Agilent, Santa Clara, CA, USA) at 50 °C and a thermal conductivity detector.

3. Results and Discussion

3.1. Effect of Gas Flow Rate and Inlet Concentration

The inlet and outlet concentrations of butanol and the responses of removal efficiency are shown in Figure 2 with respect to time of operation.

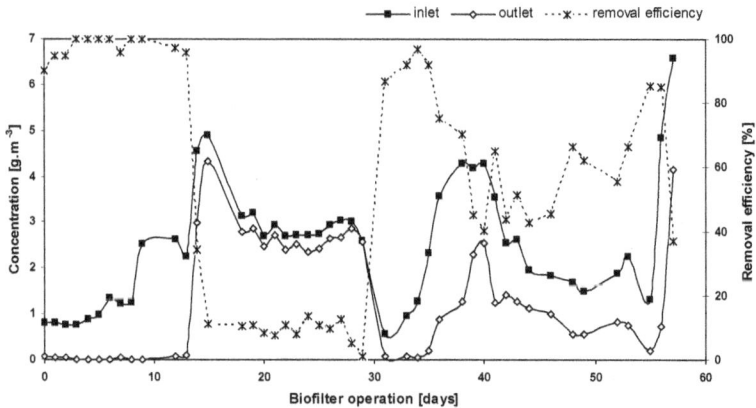

Figure 2. Inlet and outlet butanol concentration in air and removal efficiency during continuous operation of the biotrickling filter (BTF).

After inoculation, the combined effect of butanol inlet concentration and gas flow rate was investigated with respect to removal efficiency and elimination capacities. At start-up, the biotrickling filter (BTF) was run with a butanol inlet concentration of around 0.8 g m^{-3} and an empty bed residence time (EBRT) of 124 s. From Figure 2, it is clear that the BTF had a rapid development of activity after initial start-up. Removal efficiencies of more than 94% were reached almost immediately, corresponding to an elimination capacity of approximately 20 g m^{-3} h^{-1}. During the start-up phase, butanol removal might have been due to adsorption on the perlite as the initial biomass concentration was low, but as a biofilm formed over the perlite particles, the removal mechanism must have switched to biodegradation. A similar observation has been reported by Arulneyam et al. [9].

After the sixth day, increasing the inlet concentration to 2.6 g m^{-3} and decreasing the EBRT to 90 s still resulted in removal efficiencies of at least 96% being maintained for 7 days, equivalent to an elimination capacity of approximately 100 g m^{-3} h^{-1}.

The concentrations were subsequently increased to determine the maximum performance of the BTF, and as the butanol concentration exceeded 4 g m^{-3}, removal efficiency rapidly decreased. Over the period between day 15 and 29 of operation, the removal efficiency dropped to values lower than 14%, indicating that the filter was overloaded and an inhibition effect of butanol loading might have occurred. Here, the corresponding substrate mass loading rates were in the range of 110 to 120 g m^{-3} h^{-1}. It is known that exposure to high contaminant loading rates can cause inhibition effects or even kill the microorganisms, and a similar decline in performance was observed in other research [15], where it was attributed to inhibition effects at high contaminant concentration as well. Not until the butanol concentration was reduced to values under 1 g m^{-3} and after 13 days of operation, was the removal efficiency restored. However, when high loading rates (110 to 120 g m^{-3} h^{-1}) were applied at a constant empty bed residence time of 90 s, the removal efficiency rapidly dropped to near 10% and then remained almost constant at this low value. Overall, for concentrations lower than 3 g m^{-3}, removal efficiencies >60% could easily be maintained, but concentrations >4 g m^{-3} led to a rapid drop in the removal efficiency from >60 to 15%.

After day 41, the EBRT was further reduced to 60 s for a period of 8 days to observe changes in performance. The removal efficiency fluctuated between 43 to 67% with an elimination capacity of around 70 g m^{-3} h^{-1}. Despite the lower butanol concentrations in this period, high gas flow rates led to higher elimination capacities but lower removal efficiency compared with the lower gas flow rate.

Finally, during the last 5 days of operation, the EBRT was increased to 90 s and the butanol inlet concentration was rapidly increased to provide data on the maximum elimination capacity, which remained at approximately 100 g m^{-3} h^{-1}.

It is noteworthy that the variation of gas flow rates led to a relatively low variation in the removal efficiency, suggesting that the gas to biofilm mass transfer was not significantly limiting. The rapid variations of the removal efficiency show the fast response of the BTF and the microorganisms to changing conditions. Since the BTF was challenged with higher loading rates of butanol, most of the time the efficiency ranged between 50 and 90%.

The elimination capacity is plotted as a function of the mass loading rate in Figure 3. Similar to most biological reactor operations, the relationship is linear (i) up to a critical substrate mass loading rate (between 80 and 100 g m^{-3} h^{-1}), after which elimination capacity approaches a maximum value of 100 g m^{-3} h^{-1} asymptotically (ii), where it remains nearly constant and independent of the mass loading rate.

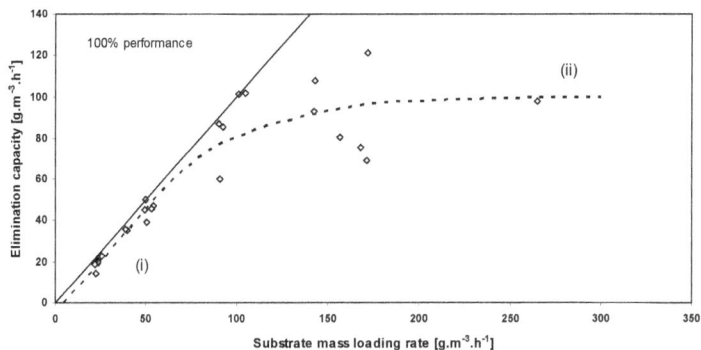

Figure 3. Effect of butanol mass loading rate on the elimination capacity of the BTF, showing (i) a linear region, and (ii) asymptotic approach to the maximum elimination capacity.

Figure 2 also shows the effect of the inlet concentration of 1-butanol on the elimination capacity and removal efficiency. Removal efficiency exceeded 80% for 1-butanol inlet concentrations up to 1.2 g m^{-3} but generally diminished with concentrations beyond this value. This range of butanol concentration can be considered as a safe operating regime. Jorio et al. [15] suggested that identifying a safe range of operation is crucial for long-term performance and can be helpful in monitoring stability and operation.

3.2. Evolution of the Pressure Drop

Nutrient supplementation was carried out twice during the operation of the BTF to maintain performance, at days 27 and 40. The additional nutrients tended to correlate with the removal efficiency, as can be observed in Figure 2. At day 27, the nutrient solution was supplied at a constant flow rate of 0.277 m^3 h^{-1}. The BTF attained a high removal efficiency of 86 to 93% between days 30 and 35, compared with efficiencies between 7 and 13% for the prior 15 days. After a while, the nutrient solution was consumed and removal efficiency decreased.

Visually, excessive accumulation of biomass on the perlite was observed as well as an increasing pressure drop after each application of nutrients (Figure 4), with the result being a partial clogging of the filter bed. If nutrient supply and concentration will, to some extent, improve the elimination capacity of a biological reactor, it will nevertheless also enhance microbial growth and initiate a pressure drop increase, as observed in the experiment. A reasonable pressure drop is crucial for optimal performance of the biotrickling filter because it directly impacts the energy required to drive air through the bioreactor, which is a substantial part of the operating costs. A long-term pressure drop under the recommended value of 700 Pa m^{-1} packing material is suggested [6], but in Figure 5 it can be seen that the pressure drop exceeded 8000 Pa m^{-1}, at which point the experiment had to be discontinued. Potential methods, such as backwashing, for removing excess biomass in biotrickling filters have been identified and discussed [8] but could not be implemented with the available equipment in this work.

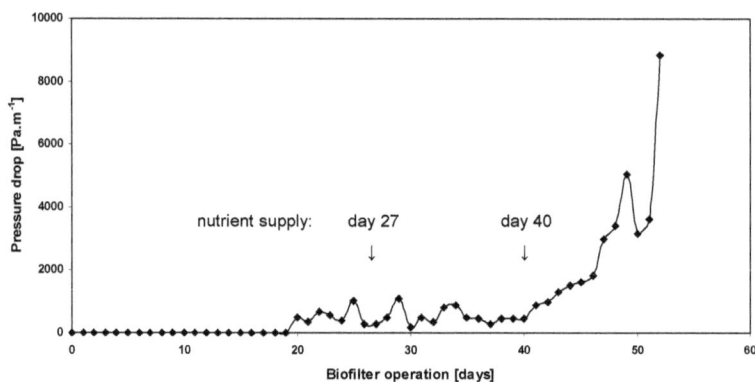

Figure 4. Profile of pressure drop across the bed height during BTF operation.

3.3. CO_2 Production

In a biological air treatment process most VOCs are aerobically degraded to water, carbon dioxide, and biomass with an approximate formula of $C_5H_7NO_2$ [27]. Hence, another method employed for monitoring biotrickling filter performance and acquiring information on the extent of mineralization is based on the measurement of CO_2 concentrations in the inlet and outlet air streams. For this purpose, the CO_2 production rate was calculated and the results are shown in Figure 5 illustrating a linear relationship between CO_2 production rate and elimination capacity, confirming previous observations [28]. If all carbon coming from butanol degradation was mineralized to CO_2, the slope observed on such a plot should be approximately 2.38 since for every mole of butanol degraded (74 g), 4 moles (176 g) of CO_2 would be produced, according to stoichiometry.

From Figure 5, the slope was estimated to be 1.09 (R^2 = 0.72), which is much lower than the stoichiometric value of 2.38, indicating that a substantial proportion of organic carbon was incorporated into various components of microbial biomass and/or used for cellular maintenance and by-product formation. Although 1-butanol is somewhat soluble in water, analysis of the recirculating water failed to detect any significant amounts in that phase.

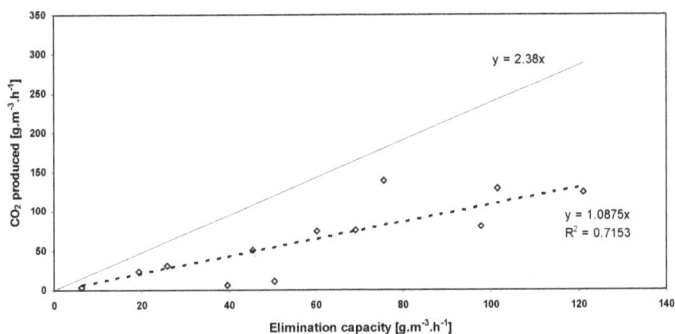

Figure 5. CO_2 produced as a function of the butanol elimination capacity, compared to the theoretical relationship (slope 2.38) with complete and stoichiometric conversion.

From the data, it can be estimated that about $46 \pm 10\%$ of the removed butanol was converted into CO_2, corresponding to a carbon recovery of 0.46 mol CO_2/mol C, which is relatively low compared to values reported by Grove et al. [16]. Assuming that the missing carbon is all converted to biomass with the formula $C_5H_7NO_2$ [16], the biomass yield would be 1.04 g biomass/g C which is likewise high compared to the suggested range of 0.17 to 0.43 g biomass/g C [16]. However, since the biofilter was operated in a biotrickling mode with pH control and nutrient supplementation, it is feasible that higher biomass growth yields were achieved in comparison to Grove et al. [16] who examined yields in a conventional biofilter (i.e., without liquid flow). In comparison, Kim and Sorial [27] measured biomass yields in biotrickling filters of 0.64 and 0.94 g biomass/g substrate for MEK and MIBK, respectively, which compares well with this work (0.67 g biomass/g butanol).

3.4. Comparisons of Performance

A literature overview on alcohol VOCs removal under similar conditions was done to compare the elimination capacities, as shown in Table 1. During operation, the elimination capacity of the filter bed varied, approximately, between 10 and a maximum of 120 g m^{-3} h^{-1}. Over the longer-term, a stable maximum elimination capacity of approximately 100 g m^{-3} h^{-1} was achieved. Results for the removal of butanol under similar conditions are relatively uncommon in the literature, but the measured elimination capacity (100 g m^{-3} h^{-1}) is comparable to those few previously reported (102 to 160 g m^{-3} h^{-1}) and for other short chain alcohol VOCs. Wang et al.'s result of 317 g m^{-3} h^{-1} stands out [22], but their air included butyl acetate and phenylacetic acid, and some potential interactions were noted when inhibitory effects were examined.

Table 1. Performance comparison between this work and selected other studies on C1 to C4 alcohols using biofilters or biotrickling filters.

Type of Process	Type of VOC	Elimination Capacity (g m^{-3} h^{-1})	Reference
Biofilter	Ethanol	53–219	[29]
Biofilter	Methanol	113	[29]
Biofilter	Ethanol	195	[9]
Biotrickling	Methanol	1916	[29]
Biofilter	n-Butanol and sec-butanol	56 and 21	[11]
Biofilter	Methanol	250	[30]
Biotrickling	Butanol	102	[17]
Biofilter	Butanol (transient)	131	[21]
Biofilter + biotrickling	Butanol (in mixture)	2	[19]
Biofilter	Butanol	162	[31]
Biotrickling	Butanol (in mixture)	317	[22]
Biotrickling	Butanol	100	This work

VOC: Volatile Organic Compound.

4. Conclusions

The laboratory-scale perlite-packed biotrickling filter was operated successfully for a period of 60 days, demonstrating effective and efficient removal from air for higher butanol concentrations than have been commonly reported (up to 4.65 g m^{-3}) with a maximum elimination capacity of 100 g m^{-3} h^{-1}. After exceeding butanol inlet concentrations of 4 g m^{-3}, the removal efficiency rapidly dropped from 90 to 15%, indicating an inhibition effect. Butanol concentrations up to approximately 1.2 g m^{-3} can be considered as a safe operating region since removal efficiencies were stable and above 80%. Carbon dioxide production was shown to correlate well with the pollutant removal. Nutrient supplementation improved the elimination capacity, but it resulted in excess biomass growth and rising pressure drop. To prevent an increase of pressure drop resulting from biomass overgrowth, a lower and controlled level of nutrient supplementation or biomass removal will need to be examined for this potential industrial application.

Acknowledgments: This work was funded by the Natural Sciences and Engineering Research Council of Canada through the Engage Grants program.

Author Contributions: The authors jointly conceived and designed the experiments. Thomas Schmidt performed the experiments, data analysis, and wrote a preliminary version of the manuscript. William A. Anderson provided technical guidance and oversight at all stages and wrote the final versions of the paper.

Conflicts of Interest: The authors declare no conflict of interest. The funding sponsors had no role in the design of the study; in the collection, analyses, or interpretation of data; in the writing of the manuscript, and in the decision to publish the results.

References

1. Bartacek, J.; Kennes, C.; Lens, P.N.L. *Biotechniques for Air Pollution Control*; Taylor & Francis Group: Abingdon, UK, 2010.
2. Cooper, C.D.; Alley, F.C. *Air Pollution Control: A Design Approach*, 3rd ed.; Waveland Press Inc.: Long Grove, IL, USA, 2002.
3. Borwankar, D.S.; Anderson, W.A.; Fowler, M.W. A technology assessment tool for evaluation of VOC abatement technologies from solvent based industrial coating operations. In *Air Quality—New Perspective*; Badilla, G.L., Valdez, B., Schorr, M., Eds.; InTech Open: Rijeka, Croatia, 2016.
4. Iranpour, R.; Cox, H.H.; Deshusses, M.A.; Schroeder, E.D. Literature review of air pollution control biofilters and biotrickling filters for odor and volatile organic compound removal. *Environ. Prog. Sustain. Energy* **2005**, *24*, 254–267. [CrossRef]
5. Delhoménie, M.C.; Heitz, M. Biofiltration of air: A review. *Crit. Rev. Biotechnol.* **2005**, *25*, 53–72. [CrossRef] [PubMed]
6. Mudliar, S.; Giri, B.; Padoley, K.; Satpute, D.; Dixit, R.; Bhatt, P.; Pandey, R.; Juwarkar, A.; Vaidya, A. Bioreactors for treatment of VOCs and odours—A review. *J. Environ. Manag.* **2010**, *91*, 1039–1054. [CrossRef] [PubMed]
7. Kennes, C.; Rene, E.R.; Veiga, M.C. Bioprocesses for air pollution control. *J. Chem. Technol. Biotechnol.* **2009**, *84*, 1419–1436. [CrossRef]
8. Schiavon, M.; Ragazzi, M.; Rada, E.C.; Torretta, V. Air pollution control through biotrickling filters: A review considering operational aspects and expected performance. *Crit. Rev. Biotechnol.* **2016**, *36*, 1143–1155. [CrossRef] [PubMed]
9. Arulneyam, D.; Swaminathan, T. Biodegradation of ethanol vapour in a biofilter. *Bioprocess Eng.* **2000**, *22*, 63–67. [CrossRef]
10. Chan, W.C.; Lin, Y.S. Compounds interaction on the biodegradation of butanol mixture in a biofilter. *Bioresour. Technol.* **2010**, *101*, 4234–4237. [CrossRef] [PubMed]
11. Chan, W.C.; Lai, Y.Z. Biodegradation kinetic behaviours of n-butyl alcohol and sec-butyl alcohol in a composite bead biofilter. *Process Biochem.* **2009**, *44*, 593–596. [CrossRef]
12. Kwon, H.M.; Yeom, S.H. Design of a Biofilter Packed with Crab Shell and Operation of the Biofilter Fed with Leaf Mold Solution as a Nutrient. *Biotechnol. Bioprocess Eng.* **2009**, *14*, 248–255. [CrossRef]

13. Kam, S.K.; Kim, J.K.; Lee, M.G. Removal characteristics of mixed gas of ethyl acetate and 2-butanol by a biofilter packed with Jeju scoria. *Korean J. Chem. Eng.* **2011**, *28*, 1019–1022. [CrossRef]

14. Kibazohi, O.; Yun, S.I.; Anderson, W.A. Removal of hexane in biofilters packed with perlite and a peat-perlite mixture. *World J. Microbiol. Biotechnol.* **2004**, *20*, 337–343. [CrossRef]

15. Jorio, H.; Bibeau, L.; Heitz, M. Biofiltration of Air Contaminated by Styrene: Effect of Nitrogen Supply, Gas Flow Rate, and Inlet Concentration. *Environ. Sci. Technol.* **2000**, *34*, 1764–1771. [CrossRef]

16. Grove, J.A.; Zhang, H.; Anderson, W.A.; Moo-Young, M. Estimation of Carbon Recovery and Biomass Yield in the Biofiltration of Octane. *Environ. Eng. Sci.* **2009**, *26*, 1497–1502. [CrossRef]

17. Heinze, U.; Friedrich, C.G. Respiratory activity of biofilms: Measurement and its significance for the elimination of n-butanol from waste gas. *Appl. Microbiol. Biotechnol.* **1997**, *48*, 411–416. [CrossRef]

18. Fitch, M.W.; England, E.; Zhang, B. 1-Butanol removal from a contaminated airstream under continuous and diurnal loading conditions. *J. Air Waste Manag. Assoc.* **2002**, *52*, 1288–1297. [CrossRef] [PubMed]

19. Lee, S.H.; Li, C.; Heber, A.J.; Ni, J.; Huang, H. Biofiltration of a mixture of ethylene, ammonia, n-butanol, and acetone gases. *Bioresour. Technol.* **2013**, *127*, 366–377. [CrossRef] [PubMed]

20. Ondarts, M.; Hort, C.; Sochard, S.; Platel, V.; Moynault, L.; Seby, F. Evaluation of compost and a mixture of compost and activated carbon as biofilter media for the treatment of indoor air pollution. *Environ. Technol.* **2012**, *33*, 273–284. [CrossRef] [PubMed]

21. Feizi, F.; Nasernejad, B.; Zamir, S.M. Effect of operating temperature on transient behaviour of a biofilter treating waste-air containing n-butanol vapor during intermittent loading. *Environ. Technol.* **2016**, *37*, 1179–1187. [CrossRef] [PubMed]

22. Wang, Q.H.; Zhang, L.; Tian, S.; Sun, P.T.-C.; Xie, W. A pilot-study on treatment of a waste gas containing butyl acetate, n-butyl alcohol and phenylacetic acid from pharmaceutical factory by bio-trickling filter. *Biochem. Eng. J.* **2007**, *37*, 42–48. [CrossRef]

23. Liu, S.; Qureshi, N. How microbes tolerate ethanol and butanol. *New Biotechnol.* **2009**, *26*, 117–121. [CrossRef] [PubMed]

24. Kennes, C.; Veiga, M.C. *Bioreactors for Waste Gas Treatment*; Kluwer Academic Publishers: Dordrecht, The Netherlands, 2001.

25. Estrada, J.M.; Quijano, G.; Lebrero, R.; Munoz, R. Step-feed biofiltration: A low cost alternative configuration for off-gas treatment. *Water Res.* **2013**, *47*, 4312–4321. [CrossRef] [PubMed]

26. Sorial, G.A.; Smith, F.L.; Suidan, M.T.; Pandit, A.; Biswas, P.; Brenner, R.C. Evaluation of trickle-bed air biofilter performance for styrene removal. *Water Res.* **1998**, *32*, 1593–1603. [CrossRef]

27. Kim, D.; Sorial, G.A. Nitrogen utilization and biomass yield in trickle bed air biofilters. *J. Hazard. Mater.* **2010**, *182*, 358–362. [CrossRef] [PubMed]

28. Rene, E.R.; López, M.E.; Veiga, M.C.; Kennes, C. Performance of a fungal monolith bioreactor for the removal of styrene from polluted air. *Bioresour. Technol.* **2010**, *101*, 2608–2615. [CrossRef] [PubMed]

29. Kennes, C.; Thalasso, F. Waste Gas Biotreatment Technology. *J. Chem. Technol. Biotechnol.* **1998**, *72*, 303–319. [CrossRef]

30. Mohseni, M.; Allen, D.G. Biofiltration of mixtures of hydrophilic and hydrophobic volatile organic compounds. *Chem. Eng. Sci.* **2000**, *55*, 1545–1558. [CrossRef]

31. Eshraghi, M.; Parnian, P.; Zamir, S.M.; Halladj, R. Biofiltration of n-butanol vapor at different operating temperatures: Experimental study and mathematical modeling. *Int. Biodeterior. Biodegrad.* **2017**, *119*, 361–367. [CrossRef]

MDPI AG

St. Alban-Anlage 66

4052 Basel, Switzerland

Tel. +41 61 683 77 34

Fax +41 61 302 89 18

http://www.mdpi.com

Environments Editorial Office

E-mail: environments@mdpi.com

http://www.mdpi.com/journal/environments